PREFACE

The aim of this monograph is to provide an introduction to
the use of quantitative, computer-based models in agriculture
in as simple and understandable a form as possible. Such
models are an important way of picturing the whole production
systems which are the farmer's basic management units and in
which research results ultimately must be applied. The
modelling activity is helpful in assessing priorities within
ongoing research projects, by setting the available information in context and pinpointing gaps and inconsistencies.

Computer modelling was earlier developed and has been
more widely applied in engineering and the physical sciences
than in agriculture; but the agriculturalist often finds difficulty in identifying with the unfamiliar subject-matter and
examples of texts addressed to workers in those fields. Nor
is he often inclined to take the alternative route into the
subject via the apparently more theoretical accounts from the
standpoint of the professional mathematician. There is a need
for considerable mathematical expertise in some specialized
aspects of modelling, especially in developing and testing new
techniques; but this is no reason for denying access to the
wide range of applications which make no such demands and where
biologists and agriculturalists can make their own, unique contribution to this multidisciplinary activity from their knowledge of the basic components and processes.

The potential for this contribution is especially great
where it is possible to harness information on the biological
basis of agricultural production more or less directly, in
mechanistic models which simulate the behaviour of system
components and processes in terms which are familiar to the
biologist: hence the emphasis on this type of model formulation
as a natural and productive way of assembling and applying
research results.

It is to be expected, perhaps, that at this stage in the
development of a relatively new area of activity there will
be some biases and inequalities and the chief one reflected

here is towards the biology of production, with less attention
to management and sociological aspects generally. In this
respect I can only plead the obvious need to redress the
balance in the future by developing a more comprehensive
treatment of 'people in agriculture', to complement that of
the plants and animals we seek to manipulate.

Hurley

November 1978   N.R.B.

Computer modelling in agriculture

Computer modelling in agriculture

BY

N. R. BROCKINGTON

CLARENDON PRESS
1979

Oxford University Press, Walton Street, Oxford OX2 6DP

OXFORD   LONDON   GLASGOW
NEW YORK   TORONTO   MELBOURNE   WELLINGTON
IBADAN   NAIROBI   DAR ES SALAAM   LUSAKA   CAPE TOWN
KUALA LUMPUR   SINGAPORE   JAKARTA   HONG KONG   TOKYO
DELHI   BOMBAY   CALCUTTA   MADRAS   KARACHI

© OXFORD UNIVERSITY PRESS 1979

British Library Cataloguing in Publication Data

Brockington, N R
  Computer modelling in agriculture.
  1. Agriculture—Mathematical models
  2. Agriculture—Data processing
  I. Title
  630′.1′84   S494.5.M3   79-40342

  ISBN 0-19-854523-1

Printed in Great Britain by
Thomson Litho Ltd, East Kilbride, Scotland

ACKNOWLEDGEMENTS

It is a pleasure to record my thanks to Professor E.K. Woodford, formerly Director of the Grassland Research Institute, for his permission and active encouragement to undertake this work. My sincere thanks are also due to Professor C.R.W. Spedding, Head of the Department of Agriculture and Horticulture of the University of Reading, for his support and wise counsel throughout, and to the staff of the Oxford University Press for their kindness and invaluable help in the editing and final preparation of the book.

My colleague Dr. P.R. Edelsten designed the FORTRAN programs given in Appendix II and these were modified for publication by Mrs. H.D. Neal and Mr. A. Windram. Dr. Edelsten and Dr. P.A. Geisler made helpful suggestions on the text and reference list. Despite the help of these and other colleagues, the responsibility for errors and omissions remains my own, of course.

Grateful thanks go to Mrs. W.A. Hassell for her rapid and accurate typing of drafts and to Mr. B.D. Hudson for his skilful preparation of the illustrations.

To my wife, Joan, I owe a special acknowledgement for her unfailing help and support.

CONTENTS

1
INTRODUCTION

1.1. SYSTEMS LARGE AND SMALL IN AGRICULTURE

In biology and agriculture, as in other subjects, we see things differently according to where we are standing and our purpose and prejudices. Between the constituent farms over a whole agricultural region and the individual components within a single plant or animal cell lies a whole range of possible views, forming a graded series in terms of the area surveyed.

Starting at the microscopic end of the range, successively larger views necessarily involve more aggregation, or less detail, within the field of interest. For some purposes, cells may be viewed not as entities in their own right but simply as constituent parts of organs or whole organisms; whole plants and animals may be seen not as individuals but as parts of populations, the crops and flocks and herds of the farmer. A complete farm commonly contains a number of plant and animal populations and those populations are integrated with the buildings and implements and other physical resources to constitute the whole-farm unit; and so on, up the scale, to the national and world agricultural scenes. Throughout this series of views there is a natural balance between the size of field and the degree of detail which it is sensible to consider within it, so that the relative position remains roughly constant. If we make a conscious effort to ignore the absolute size of a field and the specific identities of the objects within it, there is a remarkable similarity between them all and their common features may be represented diagrammatically as in Fig.1.1(a).

In each case we are looking at a number of constituent parts or 'components', which are linked together for a common purpose or function. In other words we are looking at a 'system', in which the individual parts operate together as an entity or 'whole', rather than as a simple collection of bits and pieces with no connections between them.

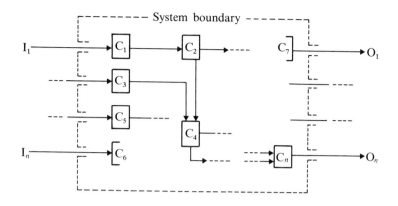

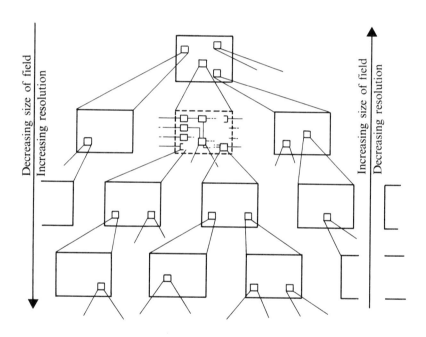

Fig.1.1. (a) Generalized diagram of a biological/agricultural system, with I_{1-n} inputs, C_{1-n} components, and O_{1-n} outputs. (b) The hierarchy of systems in biology and agriculture.

Further, as indicated in Fig.1.1(b), if we take a single component of one system within the series it is possible to focus down on that single component and treat it as a system itself, to discern within that narrower field a set of components, much smaller, but contributing to over-

all behaviour in an exactly analogous way. Alternatively, if
we choose a wider perspective than our original system, then
that system becomes merely a single component amongst a number
of larger constituents of a bigger system. Overall, agriculture and the physical, chemical, and biological sciences
which contribute to its understanding may be seen as a hierarchical set of systems. Each layer within the hierarchy comprises systems of similar size viewed in a comparable degree
of detail. Each layer is sandwiched between an overlying
layer of larger systems and an underlying layer of smaller ones
and the layers are related such that a whole system can be
seen as a system-component in the overlying layer and a system-
component can be regarded as a whole system in the underlying
layer.

   Such a hierarchy can accommodate the considerable range
of studies within agricultural science in the wide sense and
it serves to give some perspective to the wide range of purposes and objectives of those studies. Generally speaking,
any purposes and objectives must relate to the understanding
and hence prediction of system behaviour at a given level
within the hierarchy and the study and research typically
involves some investigation within the underlying layer, so
as to define the 'cause-and-effect' mechanisms contributing
to that behaviour. Additionally, an applied research project
may require that the results of the research be put into
their correct context by assessing their impact within the
overlying, more coarsely defined, layer; but the major in-
vestigatory activity remains that of looking for answers to
'How?' and 'Why?' questions in the underlying layer and that
activity is common to the majority of projects, whether they
are defined as 'applied' or not. In seeking information to
explain how systems work and especially in harnessing the
results to predict system behaviour, we are necessarily involved in constructing some form of picture or 'model' of
the system concerned. A model is required as a framework into
which the explanatory material can be slotted in an orderly
and systematic fashion and indeed as a guide to what we are
looking for, to specify which slots need to be filled in order
to make predictions of how the system will behave in varying

circumstances.

Such as the essential case for model-building in agricultural research and development projects. In the remainder
of this chapter we consider some of the general characteris-
tics of the systems to be modelled and the types of model
which are available to do the job.

1.2. AGRICULTURAL SYSTEMS AND THEIR PROPERTIES

1.2.1. *Interdependence of system components*

The central feature which distinguishes systems from 'non-
systems' is the existence of links or connections between
the constituent parts. The links are often more immediately
obvious in physical, man-made systems than in biological
examples. Commonly, it is possible to see and/or touch the
links between the components in mechanical systems (machines).
The same is true of water-supply systems and systems for
generating and distributing electricity. Notice, also, that
the word 'system' is used in such instances as part of our
everyday language, not as a specialized, jargon term. The
links and interdependence of the components in biological
systems are often less immediately apparent, although they
may have considerable scientific interest and practical
significance. In a pasture/grazing-animal system there is
an important chain of events whereby plant nutrients may be
recycled. Some of the nutrients in the herbage consumed by
the animals are voided in their excreta and a proportion can
be washed back into the soil and thus become available for
reabsorption and use by the herbage plants. Although each
stage in this cycling of mineral nutrients involves a physical movement of material, the individual links and their
overall significance are not immediately obvious.

1.2.2. *Boundaries of systems*

Both physical and biological systems are commonly classified
as 'open' or 'closed' systems. A closed system has no (significant) exchange of material with the environment in which
it operates: in other words it is self-contained; whereas an
open system is characterized by materials entering and leaving
it across the boundary. That distinction is not of special

significance in agriculture because the majority of systems
we have to deal with are necessarily treated as open ones.
Although it would be tidier and intellectually more satis-
fying to enlarge our field of view to a point where no significant transfers across the system boundary occurred; this
is seldom, if ever, possible in practice. Tracing the ori-
gin of inputs to the system and the ultimate destination of
outputs from it and adjusting the boundary accordingly is
likely to lead to the absurd conclusion that there is but
one enormous system - which it would be impossible to describe or handle. Fortunately, the modelling of open systems
is perfectly feasible in most cases. But there is one special
instance where it is unsatisfactory. If material *leaving* the
system so influences the environment, outside the defined
system boundary, that one or more *inward* flows are thereby
altered, the chain of reasoning and calculation is incomplete and we cannot derive a full description of how the
system operates. To complete the picture it is necessary
to adjust the system boundary so that the relevant part of
the environment is included within the system. This sort
of adjustment is illustrated, conceptually, in Fig.1.2 and
may be appreciated in practical terms by the following
example.

Suppose that we are concerned with the heat balance of
an animal in the field. In that situation there are important exchanges of heat between the animal and the surround-
ing atmosphere. But, generally, the effective size of the
atmospheric 'sink' for heat is such that heat passing from
the animal is effectively dissipated once it has left the
immediate micro-environment created by any layer of hair or
wool. Consequently there is no quantitatively significant
effect of such outward heat flow on the inward transfer to
its body and it is reasonable to set the system boundary
just outside the physical boundary of the animal. The situation is very different, however, if the animal should be en-
closed in a relatively small box. There may then be an ap-
preciable build-up of heat from the animal in that part of
the atmosphere inside the box and a complete picture of its
heat balance cannot be achieved unless we enlarge the system

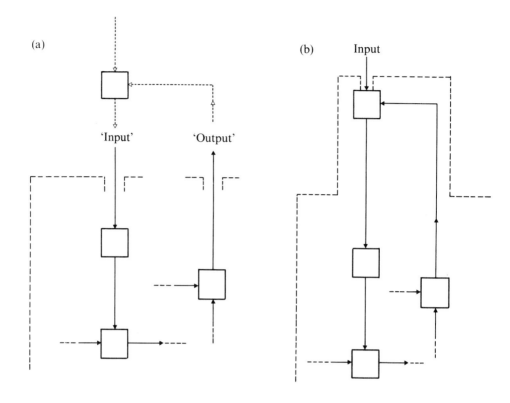

Fig.1.2. Choosing a suitable system boundary. (a) Incorrect location of
boundary - ignoring feedback loop. (b) Adjusted boundary - including
feedback loop.

boundary to include the box. Similar considerations apply to
many controlled-environment chambers in which the size and
activities of the organisms within them are incompletely com-
pensated for by the control mechanisms. That distinguished
pioneer of modelling biological systems, C.T. de Wit, has
emphasized the importance of this exception to the general
rule that open systems are to be expected and can be tolerated.
As he puts it, '... the boundary should be chosen so that the
environment influences the system, but the system itself
should not influence the environment. To achieve this goal
it is often necessary to consider a larger system than seems
necessary for the purpose' (de Wit & Goudriaan 1974).

1.2.3. *Time-dependence of biological systems*

The example of describing the heat balance in an animal's
body may serve, also, to illustrate a further important characteristic of such biological systems, their time-dependence.
In systems terminology they are 'dynamic': they change with
time and it is impossible to describe them adequately without
reference to the time dimension. If the animal we are considering is a warm-blooded or homoiothermic one, its control
mechanisms will operate so as to maintain as constant a body
temperature as possible; but the mode of operation of those
mechanisms cannot be fully appreciated unless their reactions
to varying conditions over a period of time are considered.
Because self-regulatory mechanisms like the temperature
control of homoiotherms are at least as common in biological
systems as they are in the more sophisticated man-made machine
systems, they may give a superficial impression of static behaviour. It is, however, only an apparently static condition. The fundamental character of the individuals and populations of organisms which we seek to manipulate in agriculture is based on balances between opposing changes over
time, as the organisms adjust to changes within themselves
and in the environment in which they live. As agriculturalists we have a special interest in organic growth and
the essential time-dependence of that phenomenon is perhaps
the simplest and most conclusive indication that we must be
concerned with dynamic pictures of the systems we wish to
use.

1.3. MODELS OF AGRICULTURAL SYSTEMS

1.3.1. *Understanding through simplification*

'Models have become established as a means for understanding
concepts which elude the brain's unaided ability' (Radford
1968). Models can aid in the understanding of some particular aspect(s) of a real object because they are simplifications of reality. The essential element in their construction is a degree of abstraction of particular features
of the object modelled, features chosen for their relevance
to a particular purpose. Thus the essential attributes are
presented uncluttered by all the other features which are

not essential; we may hope then to see the wood, unconfused
by the trees.

Because much follows from this basic proposition that a
model is a deliberate simplification, it must never be for-
gotten! The chief pitfall for the unwary is to be lured
into the unattainable goal of a complete, facsimile model,
perfectly representing the real object in every last detail.
If that goal were attainable in practice it would not only
contradict the definition of the word model but, more im-
portantly, it would destroy any prospect of aiding understanding.

1.3.2. *Degrees of understanding and purposes*

'Understanding' is perhaps one of the most abused terms in
current use among agricultural scientists. On the face of
it there seems little reason, beyond the strictly pedantic,
to seek to define just what we mean by the word. But it is
grossly inadequate as a statement of purposes or objectives
if it is used alone and unqualified, and since modelling
is a process of simplification for some defined purpose(s)
it is vital to specify for what the 'understanding' is to
be used. We need to probe what sorts of understanding are
required for different purposes if we are to make any real
progress in modelling (Spedding and Brockington 1976).

One may require simply to *operate* an existing agricultural production process on a push-button basis, as many
users of television sets are accustomed to do. Clearly that
is a very limited understanding indeed, but well defined and
of practical importance. A more detailed grasp of how the
system operates, often including some knowledge of its internal components, is necessary if it is malfunctioning and
needs *repair*. Even deeper and more precise understanding may
be essential to *improve* the performance of an existing system
or to *design* a new one. Throughout this wide range of purposes for which understanding can be required and for which
appropriate models may be used to aid our comprehension there
is indeed a common thread of predicting system behaviour in
some way; but the degree of understanding required for the
different sorts of prediction in different circumstances

varies greatly. We return to this argument in more detail
in Chapter 6 when discussing how to test and use models,
but the general principle is of such central importance
that it cannot be over-emphasized in this introductory discussion.

1.3.3. *Physical versus conceptual models*
Models can take many different forms but in building up
pictures of agricultural systems we are concerned exclusively with the broad class of conceptual or 'theoretical' forms,
rather than with physical models. This concentration on
paper exercises is justified by their much greater flexibility and ease of manipulation as compared with physical
representations.

1.3.4. *Stages in conceptual model-building*
In elaborating a conceptual model there is commonly a recog-
nizable progression through a number of stages. An initial
description often takes the form of a word-picture, perhaps
supported by simple diagrams and/or tables of quantitative
data. While such a description is a natural way to make a
start on the job it suffers, typically, from some ambiguity
and may well be incomplete. Especially, it is difficult
to be sure how well a verbal description meets a specific
purpose and whether there are any omissions or redundancies.

   To improve on verbal description the second stage
commonly consists of constructing a pictorial representation
of the system, to supplement or replace the written word.
The essential function of this type of picture is to describe more clearly and easily the *structure* of the system,
so that we can see not only which components are involved
in the model but also the way in which they are linked together. For that reason it is usual to employ some standard symbols for the various types of components and their
links, rather than to rely on a purely pictorial version in
a free format. Diagrams of this sort can aid in the process of selecting the essential features and omitting the
less important ones, partly because if one component is obviously required in the diagram it follows that others to

which it has important links may be required, also.

The third and final stage is represented by a mathematical or quantitative model, usually designed for implementation on the computer so as to cope with the laborious or difficult aspects of the calculations involved. This is normally the definitive form of the model because it combines the essential features of the preceding stages with the ability to calculate the consequences of all the aspects of the system as described: it is a 'working' version which summarizes the description and uses that summary to predict system behaviour. Although it may be argued that a computer model contains all and more than is found in the preceding stages and that it is the most concise and flexible formulation, this does not necessarily mean that the earlier stages are superfluous: it is often most efficient to arrive at the working model via the other forms since they constitute a natural progression.

1.4. MATHEMATICAL MODELS IN AGRICULTURE
1.4.1. Black-box and mechanistic models

There are, of course, many different forms of mathematical or computer models which have been or might be employed to describe biological and agricultural systems. Within the whole range a useful, if somewhat pragmatic first distinction that we may make is between the broad classes of input/output models and mechanistic ones (see Fig. 1.3). An input/output model is one built on the push-button philosophy: it is concerned only with what happens in terms of variation in the outputs according to changes in the inputs - not with how or why those responses come about. A mechanistic model, by contrast, is designed to depict not only *what* occurs but to describe, also, *how* the responses come about, by filling in the details of the causation-chains within the system itself. The alternative jargon term of 'black-box' model for an input/output type conjures up an accurate physical analogy in terms of a system where the internal workings are hidden from view.

Taking the example of crop response to application of a fertilizer, a black-box model would be concerned essentially

(a)

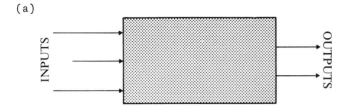

(b)

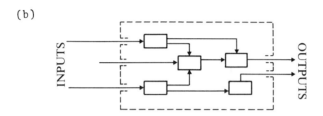

Fig.1.3. (a) Input-output or 'black-box' model. (b) Mechanistic or 'white-box' model.

with describing the change in yield resulting from varying
fertilizer input. In its simplest form this might be
summarized as a regression equation describing how the
yield response per unit of fertilizer input varies with
different levels of fertilizer. If it appeared necessary,
such a description might be elaborated to take account of
the statistical interactions between the particular fertilizer input and other inputs; but these would be concerned
simply to establish the existence of such complications as,
for instance, the dependence of the yield response on adequate levels of other plant nutrients, or on adequate rainfall to wash the fertilizer into the soil and make it available to the crop roots. By contrast, a mechanistic model of
the same overall phenomenon of crop response would attempt
to mimic some of the physical, chemical, and biological processes *within* the soil/crop system so as to describe *how* and
why the end-response in crop yield comes about. For example,
it might include an account of the water economy of the soil
so as to establish when and how the fertilizer may be leached
down the soil profile beyond the range of the crop roots, and
so on. The essential characteristic of a mechanistic model
is to mimic or imitate the relevant processes *as they occur*

in the real system, not simply to devise a convenient formula
which will predict their end-results in terms of system outputs.

   While there is a very clear distinction between extreme
examples of black-box and mechanistic model formulations it
should be recognized that intermediate types can exist, with
some elements of both forms of construction. The occurrence
of hybrid types does not invalidate the concept in principle,
of course, or diminish its utility as a way of thinking about
the various forms of mathematical model and their applications.
There is a fundamental difference between simply fitting a
formula or set of formulae in a black-box model so that they
correspond as closely as possible to the numerical values of
the inputs and outputs of a system and searching for formulae
in which the terms and parameters have a physical or biological
meaning in a mechanistic model. The black-box model is concerned simply with overall goodness of fit, irrespective of
the correspondence between the individual components of the
equations and the internal components of the real system. With
the absence of any constraint on the formulae chosen except
in terms fitting a measured set of inputs and outputs, the
chances are negligible that a black-box model will have
'biological meaning'. Thus, to the biologist, the formulae
are inevitably somewhat artificial and cannot inspire confidence by detailed correspondence with the elements which
he knows are present in the real system. Against this
apparent lack of biological realism must be set the generally
greater difficult in obtaining a comparable degree of accuracy if the larger and more complicated task of devising a
mechanistic model is attempted, with its greater number of
components and relationships to be accounted for in the equations. Intuitively, one expects that within the range of
values of inputs and outputs from which it was devised, a
black-box model will provide better predictions than a mechanistic model of the same system. On the other hand, if biological realism is an important consideration, or if extrapolation outside the measured range of input/output relations
is required, then a mechanistic model may be expected to perform better as a predictive tool because there is less like-

lihood that its behaviour will be upset by 'unexpected' deviations from reality resulting from its artifical structure.
As is generally the case in all forms of mathematical modelling, the contrasts between the expected performance of the
broad classes of black-box and mechanistic models are of
significance only in relation to quite specific, well-thought-
out purposes: models are not absolutely 'good' or 'bad', only
more or less well-fitted to particular purposes.

It is also important to appreciate that the distinction between black-box and mechanistic models is only a
relative one and depends on establishing a *point of reference*
within the hierarchical range of systems in biology and
agriculture. Since it is possible, by successively redefining
one's objectives, to include larger or smaller fields of
interest and so move up or down the hierarchy of systems, it
follows that one man's 'mechanism' is another man's 'black-box'.
It may well be, even, that an individual investigator has such
different objectives on different occasions that he can come
to describe the same object, quite legitimately, by those two
terms. Again, this does not invalidate the distinction in
principle or diminish its practical utility, provided that the
terms are qualified by reference to a particular set of objectives which serve to establish at what level within the
hierarchy one is operating. At any given level of system size
there is a real choice between a black-box view or a mechanistic one. In the former case the whole system is regarded in
the way that a passive user may treat his television set; in
the latter the attitudes of the repair-man or designer pre-
dominate.

1.4.2. Statistical and 'synthetic' models
Whenever we face the task of describing 'direct consequences'
when an event A, leads to a consequence B, or when a given
level or intensity of treatment X, leads to a given response
Y, that description in mathematical terms is a relatively
straightforward task. We are looking simply for the least
complicated equation which will fit those facts, which will
most economically summarize our verbal statement in mathematical symbols. It is seldom the case, however, when we are

concerned with biological or agricultural phenomena, that we
can make such unequivocally certain statements as those given
above. The usual situation is one where we need to hedge the
statements about with qualifications, e.g. the mean or average response Y to a particular level of treatment X is such
and such, but in a certain percentage of cases we may expect
a response lower or higher than Y by Q units, and so on. In
other words, we have to cope with *biological variation* and
we have to be satisfied with more approximate statements, both
because of that inherent variation in the material we are
dealing with and often, also, because our recordings and
measurements are not entirely accurate. To cope with such
variation and uncertainty the branch of mathematics commonly
known as 'statistics' has been developed, so that our mathematical statements can be provided with the best, most precise,
qualifications relating to the variation and uncertainty.

The methods and models of statistics are a universal
requirement in quantitative description in biology and agriculture because the variation and uncertainty are universal.
In describing systems in a quantitative way we have the basic
choice of whether to treat a particular example as a black-
box or whether to attempt to describe its internal structure.
If the decision is to treat it as a black-box then statistical models will normally be used directly to do so, e.g.
one of the family of 'regression' models to describe how one
or more of the outputs from it vary in response to changes in
one or more of the inputs. If a mechanistic picture is
chosen then what we are saying, effectively, is that instead
of treating it as one entity for our purposes we are looking
at it as made up of an assemblage of components and their
interrelationships, each of which is to be investigated in-
dividually, and the results then used to synthesize a picture
of the whole. Analysing and summarizing the results from each
of those individual investigations will normally be done
using statistical techniques, just as we might do with the
whole system when treating it as a black-box. In other words,
we have chosen to push the analysis to a greater degree of
detail and instead of one, relatively large, black box we
are now concerned with the system components as a series of

smaller black boxes.

There is, however, an additional task in a mechanistic view of an agricultural system: a description of the whole system is to be synthesized by combining the results of investigating the system components and their links within the system. Traditional research methods and procedures have concentrated to a considerable extent on the analysis phase, breaking down a system into its components, or even sub-components, and the study of those components as individual entities. But in recent years there has been increasing recognition of the importance of going on to employ the results of analysis to produce combined or *synthetic* views of whole systems (de Wit, Brouwer, and Penning de Vries 1970; Jeffers 1974). Especially in applied research, it has been realized that it is often inadequate and may be positively misleading to present the farmer or his adviser only with the bare results of analytical research on individual components. Unless an effort is made to put those results into the context of the production systems which are the operating units of the practical farmer the information may not be applied at all because its significance is not appreciated, or if an attempt *is* made it may be only partially successful because it is not clear how best to modify an existing set of management practices.

Construction of synthetic views of whole biological and agricultural systems requires mathematical models that can incorporate the results of many and often varied individual investigations and integrate those results to produce a picture of how a whole system behaves. It has been rightly commented that 'complexity is often a property of the observer rather than the system' (Spedding 1975); but the fact remains that even a minimal dissection of the majority of biological and agricultural systems reveals considerable complexity. It is fortunate that coincident with the increasing interest in models to effect a synthesis of the results of analytical research there have been made available techniques using modern computers which so augment the classical mathematical methods that it is possible to handle those models relatively easily. The possibilities for harnessing the

large electronic computer to help in using mathematical models
of such complex systems were quickly appreciated by engineers,
physicists, and business managers and the biologist has
followed their lead. In particular, the power of computers to
produce approximate solutions to the sets of nonlinear, simul-
taneous differential equations, which are often required to
represent these complex, time-varying systems, has opened up
a whole new area of theoretical biology which could not be
easily or adequately explored using classical mathematics.
Classical mathematical solutions to such problems usually are
difficult and may be impossible unless unrealistic, simplifying
assumptions are made. It is to the mechanics of building and
using these synthetic models on the computer that this text is
primarily addressed. Biologists and agriculturalists have been
slower to enter this field than their colleagues in the physical
sciences; but it appears that the need to do so is now being
widely recognized. Especially, it is apparent that while such
studies may be correctly termed 'theoretical', in that they are
concerned with paper exercises rather than with the traditional
'dirty-boots' aspects of applied research and development, they
can be of direct relevance to practical farming problems
because they are an essential stage in making the results of
research usable.

Although the construction of synthetic mathematical models
in agriculture depends in the first instance on establishing a
sound base of quantitative information on individual relationships, we do not attempt to describe those methods in this
monograph: the reader is referred to the wide range of standard
treatises which are available in that subject. Our concern is
with synthetic models and the remainder of this introductory
section outlines the broad types of synthetic mathematical
modelling techniques which are available to the agricultural
biologist and of which a selection is described in more detail
in succeeding chapters.

1.4.3. *Types of synthetic models*
The basic task to be performed in all synthetic models is to
amalgamate the available information on individual components
and processes within a system so as to come up with an integrated view of its functioning as a whole entity. To that

extent such models are all fundamentally similar; but it is
convenient to divide them into a number of types according
to their purposes and the techniques which are employed to
meet those purposes.

In the first case, we may make a distinction between
simulation[1] models designed for the general purpose of imitating system behaviour and models of an *optimizing* type,
where the specific objective is to devise some 'best' recipe
for practical operation of the system. Simulation models
may be used for investigating management strategies; but it
is generally more efficient to use specially designed optimization routines when it comes to choosing the most efficient
combination of management inputs to a system. In an overall
systems study, investigation and some description of system
behaviour are an essential prelude to devising optimum management strategies. Thus, where a simulation model is used as
a convenient way of summarizing system behaviour, the simulation study will logically precede the application of optimization techniques (see Section 6.3). In any case, the selection of optimal management strategies will depend on the
availability of at least a minimal description of the system,
of an input/output form.

Since a major aspect of systems containing living components is that they are time-dependent or *dynamic*, this
feature is important in the broad class of simulation models
which are designed to imitate their behaviour. But their
dynamic nature can be modelled in two different ways. In the
so-called *continuous* simulation models time is incremented in
small steps so that, effectively, they mimic the smooth changes
in system components that are characteristic of many biological processes. In *discrete* formulations the emphasis is
on discontinuous, abrupt changes of state and instead of
progressing through time in regular, small steps they are
designed to shift directly from one event to another, at

---

[1] The reader should be aware that the term 'simulation' has been used in a
variety of ways in the literature on physical and biological systems,
ranging from the broad usage of simply imitating system behaviour, adopted
here, to very specific definitions such as the implementation on the computer of a particular type of model designed to mimic dynamic systems by
numerical integration techniques.

whatever intervals those events may occur within the time
framework. An alternative name for the discrete simulations
is *event-oriented*, emphasizing their concentration on the
changes themselves, rather than on the time scale in which
they occur. As with the distinction between simulation and
optimizing models, there is not a rigid division between
continuous and discrete formulations in terms of how they
are used. It is quite feasible to accommodate discontinuities
within a simulation of a continuous type and this may be the
preferable way of coping with the common situation in biology
where there is a majority of processes which are smoothly
changing but there is also a proportion of discrete events.
In Chapters 2, 3 and 4 we are concerned with continuous-type
models, whether the *systems* described contain discrete events
or not; in Chapter 5 we examine the use of models specially
designed to deal with discrete changes.

The inherent variability of most biological phenomena
suggests that in assembling models of agricultural systems
based on the use of plants and animals particular attention
should be given to imitating that variability. If we have
information indicating the variation in behaviour of individual parts of a system it is possible to treat those parts
as *stochastic* variables, allowing each to vary in a random
manner consistent with the estimated statistical form of
its variability. In that way the whole-system model may
express in its behaviour an appropriate degree of unpre-
dictability, with a range of possible outcomes for a given
set of external and internal circumstances. Such stochastic
model formulations may be contrasted with *deterministic*
models, in which there is only one, predetermined consequence
of a given set of controlling conditions.

Given the widespread occurrence of variability in response of biological processes to controlling factors it is
perhaps a little surprising that deterministic models appear
to have dominated the simulation modelling scene in biology
and agriculture over the last few years. One reason may be
that in the early stages of using synthetic models in these
subjects it is commonly the case that the data available
for model construction are not as complete or reliable as

would be desired. Consequently it may be inappropriate to
attempt to build apparently sophisticated models, including
stochastic treatment of the constituent processes, when the
slender foundation of quantitative data is inadequate for the
purpose. Recognizing the limitations of the available data,
the questions posed in initial modelling work may be better
confined to elucidating the directions of system response
and their approximate order of magnitude rather than attempting precise, detailed predictions. For such purposes deterministic formulations may suffice and may be preferred on the
grounds of simplicity (Innis 1975). At the same time, there
can be no doubt that in certain circumstances an account of
variability can be an essential element in simulating system
behaviour; we return to this topic in Chapters 3 and 5.

An outline classification of the main types of model considered above is given in Fig.1.4.

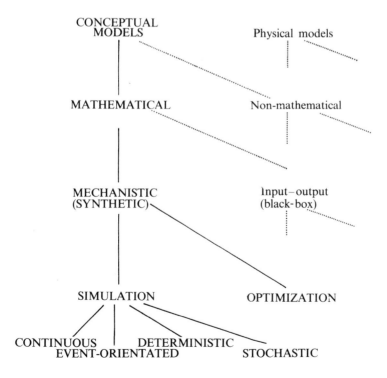

Fig.1.4. Outline classification of models in agricultural systems.

2

CONSTRUCTING DIAGRAMS FOR DYNAMIC SYSTEMS MODELS

We noted (Section 1.3.4) that in mathematical modelling of agricultural systems it is often helpful to translate the initial, word description into a *system diagram*, as an intermediate stage in progressing to the final model in the form of a computer program. This is especially so for models of time-dependent, dynamic systems, where a carefully thought out system diagram can provide an invaluable framework in which to elaborate the computer program. For the beginner in systems modelling these diagrams can help to make the not inconsiderable mental leap from a word-picture to the description in terms of mathematical equations; even to the initiated there are the advantages of a concise, ready-reference picture of system structure for use while programming and to help in describing and discussing the model with others. For these reasons, we devote this chapter to an exposition of a number of examples of such diagrams.

To avoid any possible confusion amongst those who may be familiar with the use of 'flow charts', as a pictorial representation of the exact sequence of all the arithmetical and logical steps to be executed in a computer program, we are *not* concerned here with such blueprints for the details of the calculations. We are considering the major components within the system and how they are linked together. Such a diagram of overall system structure does have important consequences for how the computer program can best be elaborated to mimic system behaviour; but it does not necessarily prescribe the precise form and sequence of the individual steps in the calculations.

It may be as well to point out, also, that we are concerned with only one of the range of types of system diagrams. These can vary considerably in relation to the purposes for which they are designed and the conventions adopted. The reader is referred to Spedding (1975) for a discussion of the general principles involved, with illustrations of various types.

2.1. AN EXAMPLE OF A MAN-MADE SYSTEM: DOMESTIC WATER SUPPLY

We consider, first, a simple non-biological system to illustrate the methodology involved.

Figure 2.1(a) is a semi-pictorial diagram of a system
for supplying water for domestic use. It serves to meet some

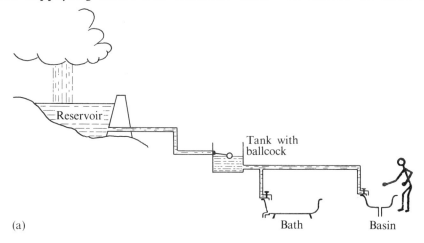

(a)

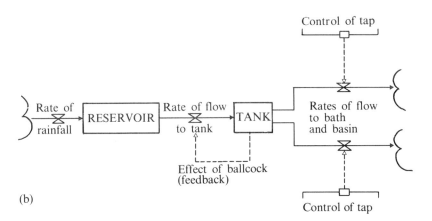

(b)

Fig.2.1. A domestic water-supply system. (a) Semi-pictorial diagram of
the system. (b) Flow diagram of the system using conventional symbols.

of the requirements for a working diagram of the system;
showing, for example, not only what components are contained
in the system but indicating, also, how they are arranged in
relation to each other. The main drawback is an element of
vagueness, largely because of the absence of any rules or
conventions for its construction. The diagram is partly

dependent on personal choice and artistic ability, and different individuals may produce different pictures of the same
system.

Experience has shown that it is preferable to adopt a
series of conventional symbols for the various classes of
components which are common to all time-varying systems. In
this text we use, with minor modifications, the conventions
of Forrester (1961). Although his approach was evolved in
order to deal with industrial and business systems, it has
proved easily adaptable for biological systems and has gained
widespread acceptance in that field.

The basis of this type of diagram is the representation
of changes in the state of the system through time by reference to the amounts or 'levels' of physical materials at
various points within the system. The locations at which the
varying amounts of material in the system are to be described
are indicated in the diagram by square or rectangular *boxes* or
compartments. Physical movements of material into and out of
the boxes are shown by solid arrows, usually with the addition
of a valve-like symbol, thus, ⌿ to signify that the movements, or *flows*, are not unlimited, but are constrained or
controlled in some way. An arrow leading to a particular
box implies a flow of material to that box, adding to its
contents; conversely, an arrow pointing away from the box
shows that material is subtracted from it.

Figure 2.1(b) uses the Forrester-type symbols to describe the water supply system and contains two boxes, one to
represent the amount of water in the reservoir and one for
the water in the house tank. There are four solid arrows to
show the flows of water into and out of the reservoir and
the tank. Clearly, in this example the fluctuations in the
level of water in those two locations over time can be accounted for by the additions and subtractions of the four flows.[1]

----

[1] The major concern with physical flows of material has led to diagrams
of this form being termed *flow diagrams*. Whilst this is an accurate
reflection of their form and function it is unfortunate that the term
may be easily confused with the *flow chart* of computer programming par-
lance (see above). Since the term is widely used, however, and has
no adequately descriptive alternative, we adopt it here.

In addition to the amounts and movements of materials denoted by boxes and arrows, there are some other classes of system elements which are represented by standard symbols. A complete description of a dynamic system involves the recognition that there may be links of two sorts between system components. The first form of connection is the physical one involving actual transference of material; this we have seen is represented by a solid arrow. The second type of link is one in which there is no physical movement involved, but where one component may be regarded, nevertheless, as influencing or controlling the behaviour of another. Such controls may be regarded as transfers of *information* rather than materials and are represented in the diagram by arrows with broken lines. Because the system is described in terms of the fluctuations of the amounts of material in the boxes and those fluctuations are a direct consequence of the physical flows, it turns out that the only points at which control of system operation is possible are the physical flows. Consequently, the broken arrows invariably are directed at the physical flows and the valve-symbol on each solid arrow provides a convenient focus at which to direct the one or more *information flows* symbolizing its controls. The controls on the rates of physical flow shown in Fig.2.1(b) are the automatic, feedback, effect of the ballcock in regulating water flowing into the house tank from the reservoir and the control, by the user, of the taps delivering water to the bath and the basin.

The remaining symbol used in this example is an irregularly shaped one denoting *sources* and *sinks*. This is employed to indicate the *source* of water entering the reservoir, i.e. it is equivalent to the rain-clouds sketched in the pictorial version. It is used, also, to show the *destination* of the water flowing from the household taps. The use of the symbol is to replace that indicating a compartment where such a compartment may be regarded as being outside the boundary of the system modelled; in other words, it shows that we recognize that a compartment exists at the point shown, but are not concerned to calculate its contents because such a calculation is not required as part of the description of the

system. As discussed in Section 1.2.2, the drawing of a legitimate system-boundary for a model of a system needs to be
carefully considered in relation to feedback loops and this
question normally arises in the choice of representing a
particular location by a (quantified) compartment, or by an
(undefined) source or sink. In the case of the water supply
system, it appears reasonable to regard the source of the
rainfall feeding into the reservoir as independent of any-
thing within the system or the outputs from it. Hence the
rain-clouds need not be included in the model; although the
rate of *arrival* of rain at the reservoir will require to be
specified, of course. Similarly, the water leaving the system from the bath and basin is assumed to be drained away independently. However, it would be obviously invalid, for
example, to postulate a boundary which cut across the link
between the reservoir and the house-tank such that the operation of the feedback control by the ballcock was incompletely
described.

2.2. A SIMPLE BIOLOGICAL SYSTEM: RELATIVE GROWTH

Having seen how the methodology works out in the familiar
context of a man-made system, we consider how it may be applied to a simple biological example. This example is one of
the simplest possible views of the growth of a plant or animal:
assuming that its rate of growth is dependent only on the size
it has already attained at any point in time. In that way, an
increment of growth is calculated by multiplying a constant
proportionality factor, the so-called relative growth rate,
by the volume or weight etc. of the organism, according to the
units employed to represent its growth. Thus if an animal
weighs 20 kg at a given time and has a relative growth rate
of 0·05 per day, the increment over the next day will be
20 kg × 0·05 = 1 kg.

Such a simple assumption is unlikely to be valid for any
organism for any length of time, but it may serve as a useful introduction to constructing flow diagrams for biological
systems. The basic form of Fig.2.2 is one box, with one
solid arrow feeding into it. The box represents the chosen
attribute of growth, weight in this instance, while the arrow

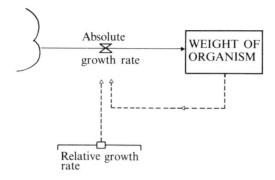

Fig.2.2. Relative growth of an organism.

indicates the rate at which the weight is increased, i.e. the growth rate. Compared with the water supply system, it is perhaps less immediately obvious that the growth process can be represented in this way. The physical flows of water through pipes are relatively easily translated into a flow-diagram form; but, in fact, the phenomenon of biological growth can be visualized as an exactly analogous process, with the entry of nutrients which add to the weight of the organism similarly represented as a physical flow.

Since the assumption has been made that it is only the weight of the organism which controls its growth rate, the supply of nutrients must be taken as unlimited and can be appropriately represented as an undefined source, rather than a finite compartment.

To complete the diagram it is necessary to show the controls on the growth rate, using the broken-arrow symbol. One of these factors, the proportionality constant, is an item of data supplied as an input to the model; we use the symbol —O— to represent such information inputs. The second control on the rate of growth is the weight of the organism; and the broken arrow leading from the box to the valve symbol represents the transfer of that information.

2.3. A PHYSIOLOGICAL MODEL: CARBON METABOLISM IN A GREEN PLANT

The model structure represented in Fig.2.3, while not so simple as the previous example, contains some fairly gross assumptions which will be readily apparent to plant physiologists. Remember, however, that the validity of those assumptions may be judged realistically only against a precise

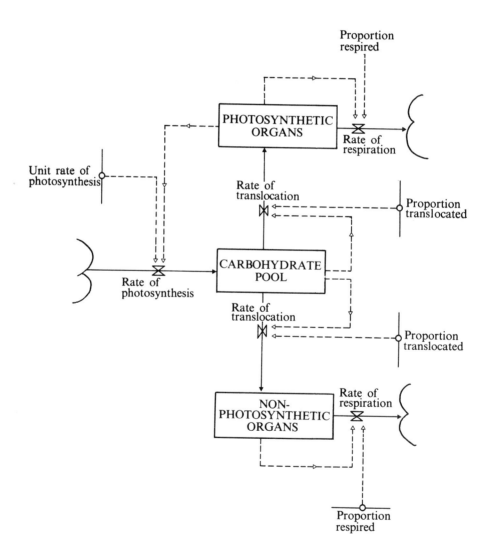

Fig.2.3. Carbon metabolism in a green plant.

statement of the objective for such a model (Section 1.3.2).
Here, we are not concerned to effect such a judgement, but
to illustrate how a somewhat more complicated picture of
biological function than the relative growth model can be
represented in diagram form, using the same basic symbols
and conventions a number of times.

The model is designed to imitate the major flows of
carbon compounds in a green, photosynthesizing plant. Carbon
enters the plant by diffusion as carbon dioxide and is fixed
in the process of photosynthesis to produce a labile pool of
simple carbohydrates. The entry and fixation rate is considered to be controlled only by a given, constant, rate per
unit mass of photosynthetic tissue, represented chiefly by
the leaves in most higher plants. Accordingly, two controls
on the combined entry plus fixation rate are shown in the
diagram as broken arrows. One represents the input of data
for the unit rate and the other the transfer of information
on the mass of the photosynthetic organs from the compartment
in which that parameter is calculated.

Carbohydrate from the pool is assumed to be allocated in
fixed proportions to the photosynthetic and non-photosynthetic
parts of the plant body. The two rates of transfer involved
in the allocation are controlled by appropriate data inputs
on the proportions and those proportions are used to arrive
at the absolute transfers in the light of information on the
amount of material in the pool. As part of the overall simplification of the model, no account is taken of the elements
such as nitrogen, phosphorus, and sulphur which are involved in the synthesis of plant tissue in addition to the
basic building blocks of carbohydrate units derived from
photosynthesis. Typically, those additional elements would
be expected to contribute something of the order of 5 per
cent to the organic matter in the new tissue. Net loss of
plant weight from the photosynthetic and non-photosynthetic
parts in respiration is taken to be a fixed proportional
amount, again involving appropriate data inputs and the transfer of information on current weights to the relevant rate
processes.

At the boundaries of the system, as modelled, carbon

enters the plant from an undefined source of carbon dioxide and the output of carbon dioxide from respiration is similarly transferred to unspecified sinks.

2.4. AN ANIMAL POPULATION MODEL

Finally, in this series of examples of diagram construction, we consider a biological system in which a population of individuals is described, with particular attention to the fluctuations in numbers consequent on births, deaths, and the abstraction of some individuals for sale. The animal population concerned is assumed to be managed for the production of animals for slaughter, with all the males sold at maturity except for the small number required for breeding and with a lesser proportion of mature females sold off.

Despite the focus on numbers of animals rather than on quantities of material, as in the previous examples, the same symbols and conventions for diagram construction may be used, as illustrated in Fig.2.4.

Young males and females are born according to the number of mature females and in accordance with data on the birth rates and sex ratio. No account is taken of the influence of mature males on the reproductive rate, it being assumed that an adequate number of this class is maintained in the herd, as in good commercial practice. Proportions of young animals dying before they reach maturity are specified by data inputs and it is assumed that there is no significant mortality of adults prior to sale.

The flows of animal numbers to the compartments representing the accumulated totals sold are taken to be calculated from the numbers of mature animals and from data inputs on the proportions which are to be disposed of in that way. This treatment is indicated in the diagram by appropriate information flows. The totals sold are, of course, of prime importance as a means of judging the practical performance of the system modelled. It is for this reason that the boundary of the model is assumed to include these two compartments, although there is no technical necessity to do so in terms of requiring the totals in order to complete any of the other calculations, i.e. no feedback loop would be cut

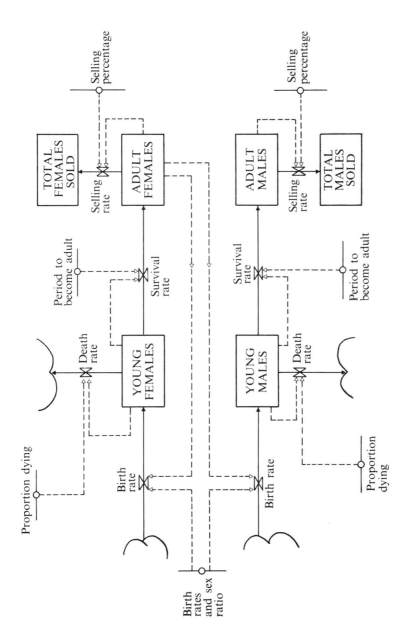

Fig.2.4. An animal population model.

across if the compartments were replaced by undefined sinks. These considerations illustrate how, in defining an appropriate boundary for the system to be modelled, a first step must be to ensure that those components which are relevant to the defined objective are included; thereafter the tentative boundary should be reviewed to ensure that any feedback loops are included in their entirety.

Again, it may be emphasized that the assumed model for this exercise in diagram construction has been considerably simplified as compared with the detailed description of a specific animal production system that might be required to meet a particular objective in practice. It may serve, nevertheless, to illustrate again the application of the principles involved.

Note that in the interests of completeness the controls on the transfer rates have been shown in the diagram in all their details and including, for example, the transfer of information from the immediately preceding compartment, which is commonly required as part of the calculation of a rate. This is good practice for the beginner, allowing for no ambiguity; but in more complicated models such donor-compartment controls are frequently omitted from a flow diagram to avoid cluttering it up with too much detail and detracting from its overall clarity. It may be necessary in some very large models to exclude all the controls on individual transfer rates and to convey that information separately from the main flow diagram. But the beginner is advised not to be tempted by such apparent short-cuts; complete mastery of this form of diagram construction forms a sound basis for understanding the process of constructing the working model in the form of a computer program. To that end, we set out below some example problems in building up diagrams of this type, with suggested solutions in Appendix I.

EXAMPLE EXERCISES

1. Construct a flow diagram for the domestic water supply in a house where a tap in the kitchen is fed directly from the mains supply, two taps in the bathroom draw

water from a tank in the roof-space and that tank also supplies water to a cistern for flushing the lavatory. Levels of water in the roof tank and in the lavatory cistern are regulated by ballcocks. (The reader may find it advantageous to draw a picture-diagram, as in Fig. 2.1(a), before attempting to construct a flow diagram.)

2.   The relative growth rate of a plant is assumed to vary with temperature. Modify the flow diagram given in Fig.2.2 to represent this situation, assuming that data on temperature and its relationship with the relative growth rate are available.

3.   The growth of a pasture is taken to be dependent on three factors: (i) the herbage present, (ii) the available soil moisture, and (iii) the temperature. Some of the herbage dies and is lost from the system and the rate of loss depends on the amount of herbage present and on the temperature. A proportion of the herbage is eaten by grazing cattle and their consumption varies according to the herbage available and the number of animals. Draw a flow diagram for this plant/animal system, assuming that data are available for soil moisture, temperature, and the number of grazing animals.

3

COMPUTER PROGRAMMING FOR DYNAMIC SYSTEMS MODELS (1): SIMPLE EXAMPLES

The computer program is normally the final, definitive form of a simulation model to imitate the behaviour of a dynamic system. In this chapter we introduce some of the tools available and their use on the computer, employing the same examples for which we constructed flow diagrams in the previous chapter.

3.1. COMPUTER LANGUAGES FOR DYNAMIC SYSTEM MODELS

Among the languages commonly available for use on modern computers there are two main groups of concern to us. First, the languages such as FORTRAN and ALGOL which are very versatile and can be used for a wide range of scientific applications, including modelling. The inherent flexibility of these languages is the result of concentrating on individual instructions of a relatively basic nature[1] and which can be used in building up programs to perform a wide range of overall tasks. Such versatility is accompanied, however, by a degree of complication in constructing and using the programs which is inevitable if the basic building blocks are to serve a wide range of purposes. The programmer must specify in considerable detail precisely what he requires the machine to do for him when using these languages.

A second group of languages comprises those which have been deliberately tailored to make it relatively simple to represent the commonly occurring features of dynamic systems. Included in this group are languages such as DYNAMO and CSMP

---

[1] Computer specialists commonly refer to languages such as FORTRAN as 'high-level' languages, regarding them as sophisticated developments because they incorporate a considerable degree of aggregation of the very simplest possible individual instructions that could be given to the computer, corresponding to the basic operations which are carried out in the machine. For our purposes, languages like FORTRAN and ALGOL represent the most detailed level of 'resolution' normally worth considering and contrasting with the even greater aggregation of commands in more specialized languages.

(Brennan, de Wit, Williams, and Quattrin 1967; Pugh 1970;
IBM 1975), which are primarily designed for *continuous* simulation models, but which can accommodate discrete formulation
of some elements. For models of an exclusively *event-
orientated* form, languages such as GPSS are available (see
Chapter 5).

The special-purpose simulation languages offer many
features which are designed to make the common operations
in programming simulation models as easy and straightforward
as possible, thus leaving the modeller free to concentrate
more of his attention on the form and use of a model, with
minimum distraction by the technicalities of computing.
The principal means by which the programming technicalities
are reduced to a minimum is by offering abbreviated, easily
learned commands for the sets of instructions that would
otherwise be required for the commonly used operations in
such models. The price that has to be paid for that simplification normally includes a reduction in overall flexibility and sometimes a degree of unnecessary complication
and artificiality in carrying out non-standard operations.
The user may expect to have a generally easier time in
learning and using these languages compared with the general
purpose types such as FORTRAN. But he must accept the oc-
casional difficulties that may arise in programming non-
standard operations which have not been foreseen by the
designers. He needs to accept, also, that in general he
will be less in control of the details of how the calculations
are handled in the computer and that this may be a somewhat
inefficient way of using the machine.

A degree of compromise between the general rigidity of
the special-purpose languages and the flexibility of general-
purpose types is available in CSMP. This language is based
on FORTRAN, with the addition of special features for
modelling applications and the user may opt to use standard
FORTRAN for sections of a program, as required.

The choice of computer languages for programming models
of biological and agricultural systems has been a keenly
debated issue on occasions over the last decade (see e.g.
Charlton 1971; Radford 1972). But it is doubtful if there

is a simple, universal answer to the problem. Important
elements in an individual's choice must be his existing skills
and experience, if any, and the type of model he wishes to
construct. An expert programmer in FORTRAN may choose to
capitalize on his existing skills and carry out all his model-
building in that language. For the complete beginner, there
can be some attraction in the shorter learning period for the
special-purpose languages and providing his requirements are
restricted to fairly simple modelling projects he is unlikely
to suffer any significant penalties in terms of complications
in the programming. Neither is it likely that he will be concerned about any possible inefficiencies in the use of computer time which may occur: computing costs normally will represent only a small item in his overall budget.

In attempting to meet as wide a range of potential requirements as possible in this text without overloading it
with excessive technicalities we present the example material
in the remainder of this chapter in CSMP and give, also, the
essential, 'core' elements of alternative programs in FORTRAN
in the main text. Complete FORTRAN programs for these examples, including the use of specially designed subroutines
for some common tasks in programming simulation models, are
explained in Appendix II. In Chapter 4 we use CSMP as the
basic explanatory aid, with an outline description of the
use of GPSS for programming event-orientated models in Chapter
5. It must be emphasized that while endeavouring to make the
example material for computer programming as self-contained
as possible, it is neither feasible nor appropriate to attempt
to give detailed and complete instructions for the use of the
computer languages concerned. The reader must refer to the
standard texts and manuals on these languages (see p.146) if
he is to become sufficiently familiar with them to put his
knowledge of the principles of model construction to practical
use. It is worth emphasizing, also, that a vital element in
acquiring the necessary practical skills in computer programming is practice, and yet more practice!

3.2. BASIC CALCULATIONS IN A DYNAMIC SIMULATION MODEL: SIMPLE NUMERICAL INTEGRATION

In progressing from the diagram stage to a computer program there are two major changes in the content of the model: an essentially qualitative picture is converted to a quantitative one and the implication of dynamic behaviour in the diagram form is realized explicitly in a 'working' version, with calculations of how the system changes over time in terms of variations in the contents of the compartments.

Calculating the changes in the levels of material in the compartments as time goes by is the basic feature of all dynamic simulation models and to illustrate how this is done we return to the example of a domestic water supply system for which we devised a flow diagram in Fig.2.1 (p.21).

Considering the compartment which represents the water tank in the house, the diagram indicates that there are three flow rates affecting the level of water: the flow *into* the tank from the reservoir and the flows *out* to the bath and the basin. If we wish to calculate the level of water 'now', at this instant of time, we may do so by reference to a previous level, say what it was last night, *adding* what has flowed in since that time from the reservoir and *subtracting* what has flowed out. Using convenient abbreviations[1], the calculation may be summarized in equation form thus:

$$CLEVTK = SLEVTK + FLOWIN - FLOWBH - FLOWBN,$$

where CLEVTK is the current level of water in the tank, SLEVTK is the starting level, FLOWIN represents the inward flow from the reservoir and FLOWBH and FLOWBN are the flows out to the bath and the basin.

There is an important limitation in using such calculations, of course, in that it is assumed that the flow rates

---

[1] Notice the use of abbreviations in this equation of a *mnemonic* form, serving to identify the computer variables with their verbal equivalents. Within the limits of the five or six alphabetic/numeric characters permitted for variable names in most computer languages, this is a good, general practice and is recommended to the beginner as greatly improving the 'readability' of computer programs.

remain constant over the whole of the time period in question. The practical dodge which is used in computer modelling to get round that constraint is to divide up the overall time period into sufficiently small subperiods such that the flow rates can be regarded as effectively constant during any one sub-period. In that way the result of *repeating* the calculations for each small time step can produce an acceptable approximation to the change in the contents of the compartment over the total time period.

In mathematical terms, the rate of change in the level of material in a compartment (x) over time (t) is properly represented by a differential equation of the general form:

$$\frac{dx}{dt} = f(x,t) \ .$$

As we shall see later (p.49), it is possible to solve such equations *analytically* for very simple models. But many applications involve sets of nonlinear differential equations which are difficult or impossible to solve analytically and in those cases it is necessary to derive approximate solutions by *numerical* methods. The simplest of these is the rectangular or Euler's method, in which an increment in x, Δx, is approximated over a small interval of time, Δt, using the difference equation:

$$\Delta x = f(x,t).\Delta t \ .$$

Adding the increment in the level to its original value gives an estimate of the new value,

$$x_{(n+1)} = x_n + \Delta x,$$

where n is the (arbitrary) starting point in time and $(n+1)$ represents the end of the 'calculation interval' or 'delta-time'. Having estimated the revised level after the lapse of one small increment in time, the process may be repeated for the next increment, and so on. Figure 3.1 shows part of such a *sequence* of calculations corresponding, for example, to the repetition of the equation for calculating the level

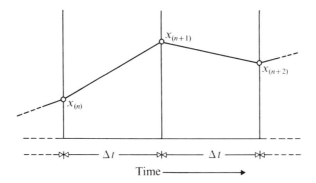

Fig.3.1. The calculation sequence in simple numerical integration: the content of the compartment is recalculated at intervals of delta-time.

of water in the tank described above. At the beginning of each calculation interval in the sequence, say time n, the information available is the *present* value for the contents of the compartment and its *present* (instantaneous) rate of change. On the basis of that information, the simplest assumption that can be made about the *future* rate of change, between time n and $(n+1)$, is that it will remain constant. That is the assumption of the rectangular method of integration and it leads to the form of approximation illustrated in Fig. 3.1, where a continuous change in the compartment contents is represented by a series of straight-line segments. Clearly, the adequacy of this form of approximation will depend on the size of the calculation interval relative to the rate of change. We return to the question of the approximations in numerical integration later (p.50); here it is sufficient to note that the majority of biological applications are well served by simple forms of numerical integration with an appropriate length of calculation interval. But to complete our introduction to the structure of computer-based simulation models we need to examine some other aspects of the calculations.

First, we can now derive a more general form of computer instruction to follow the variation in the contents of a com partment through time. While the details of the notation

will vary with the language employed, the following form may
serve to illustrate the essential elements:

$$CLEV = PLEV + DT * NRC ,$$

where CLEV is the current or 'updated' value for the level,
PLEV is the previous level, DT is the length of the calculation interval (delta-time) and NRC is the net rate of change.
For individual applications the net rate of change is ex-
panded into the appropriate number of positive and negative
elements for the compartment concerned. For example, the
level of water in the tank considered in our example would
have the form

$$CLEV = PLEV + DT * (R1-R2-R3)$$

with one positive and two negative flow components. Notice
the inclusion of delta-time as a multiplying factor (sig-
nified by an asterisk), to convert the flow-rates from a
unit-time basis to rates per calculation interval.

In applying the general formula for updating the contents of a compartment it should be noted, also, that the
first such calculation within a sequence is a special case.
In all subsequent calculations the result of the preceding
one is available as a starting point for the current operation; but this is not so for the first application at the
beginning of the overall simulation period and an appropriate numerical value for that *initial condition* must be
supplied as a *data input* to the model.

A further aspect of the calculation structure which requires some examination is the derivation of the flow rates.
In very simple cases, this may involve simply specifying
appropriate numerical values as data inputs; more often,
the 'controls' on a flow rate include the influence of other
components within the system, e.g. the control of the rate of
water entry to the tank in the domestic water system by the
feedback effect of the level of water in the tank. This exem-
plifies an important general principle, that the internal
controls on a rate of transfer are always defined by reference

to the 'state of the system', in terms of the contents of one
or more compartments, *not* by reference to other rate processes. This proposition is perhaps not immediately obvious,
partly because of the loose way in which we tend to talk about
relationships within systems. If we are considering, for
instance, the heat exchanges of an object exposed to the sun,
a statement to the effect that the rate of heat loss depends
on the heat gain from radiation may appear superficially
acceptable. In fact, there is no such consistent relationship between the rates of gain and loss; but the rate of loss
is dependent (amongst other factors) on the heat *content* of
the object. If the heat *capacity* of the object is small, the
rate of gain may appear to have an overriding effect on the
loss rate; but in other circumstances the assumption of a
direct relationship between the two rates may be seriously
inaccurate and in no case will it be a true reflection of
the *structure* of the system.

de Wit and Goudriaan (1974) have pointed out that similar considerations apply, for example, where two organisms
are dependent on a common food supply. Except in very
special circumstances, the rate of foot consumption by one
species does not *directly* affect its competitor's rate of
eating; but both rates are dependent, *inter alia*, on the
level of food available and it is only when that fact is
appreciated that we can analyse the situation reliably.

Returning to the physical example of heat exchange, we
can say that the heat content of the object represents a
'summary' at a given point in time of all the past events,
both gains and losses. Consequently it is a better guide
to the rate of loss in the immediate future than is the
current rate of input. Put in that way, the proposition
of the control of rate processes by the 'states' of compartments conforms with the sequence of calculations for updating
compartment contents that we described above. At the beginning of a calculation interval we know the current level of
material in the compartment(s) and we then project the rates
over the calculation interval on the basis of that knowledge,
including any internal controls on the rates by the compartment states.

Thus the overall sequence of calculations involves a regular *alternation* of projecting a minimum distance into the future by estimating the flow rates over the next time-increment and of summarizing the consequences of that projection by updating the levels in the compartment(s). Having summarized the consequences of a projection we are then in a position to make a projection for the following calculation interval, and so on.

So far, we have been concerned, essentially, to outline the calculations for an individual compartment plus its associated flow rates. But it is usually the case that a model of this type has more than one compartment and the question arises of the *order* of calculations for the various compartments. For a multi-compartment model we extend the rule for an individual compartment such that the sequence involves first calculating *all* the rate processes at a given point in the time sequence and then updating *all* the compartment contents. This procedure allows for all the current values of the compartments to be used, as required, in making forward projections of the rates.

Conceptually, all the rates are calculated simultaneously and the updating of all the compartments likewise is done 'in parallel', at one and the same time. In practice, the desired effect is achieved by what is termed 'semi-parallel' processing, in which it is sufficient to ensure that the whole *batch* of rate calculations is completed before dealing with the set of up-dating operations for the compartments.

Finally, in this review of the calculation structure of simulation models, we need to refer to an alternative system of nomenclature for variable-types to that which we have used up to this point. The alternative names are often used in describing the calculation structure and are sufficiently common in the literature to make it appropriate to introduce them at this point. What we have called, thus far, 'boxes' or 'compartments' are also known as *state variables* and data inputs to a model become *driving variables* or *forcing variables*. Dynamic simulation models of the type we are discussing frequently are called 'state variable models'.

3.3. PROGRAMMING THE RELATIVE GROWTH MODEL

To illustrate how the methods of computer modelling discussed
above may be applied in practice we first consider the programming of this simple view of the growth of an organism
(see Fig.2.2, p.25, for the flow diagram of the model).
Numerical values for the necessary data inputs to the model
can be taken as:

> Starting weight of the organism = 20 kg
> Relative growth rate = 0·01/day
> Total growing period = 50 days

3.3.1. A CSMP program

The first step is to supply the computer with the numerical
values which we have been given for the initial weight
of the organism and for the relative growth rate, so that
these are available when required. Using suitable mnemonic
abbreviations, an instruction in CSMP to effect those data
inputs is:

> PARAMETER IWT = 20.0, RGR = 0.01.

Having alerted the machine to the fact that numerical values
for the technical class called 'parameters' are to follow,
we state the abbreviations and their associated numerical
values.

Next the single physical flow is calculated, deriving
the absolute growth rate (AGR) as the product of the weight
of the organism (WT) and the relative growth rate:

> AGR = WT*RGR .

The special function, INTGRL, in CSMP is the overall command
for the set of calculations involved in numerical integration of one or more flow rates and the appropriate use of it
for this example is:

> WT = INTGRL(IWT,AGR) .

In this command the first term in parenthesis specifies the initial condition for the compartment and subsequent terms refer to the relevant rates. In this case only one rate is involved and we omit the optional '+' symbol which would indicate it is adding to the contents of the compartment. Use of a '-' symbol for rates which subtract from the compartment is obligatory.

Because CSMP has a range of forms of numerical integration which are available to the user, we add a further instruction which specifies that the rectangular method is to be employed.

METHOD RECT .

The above four instructions constitute the essential core of the program for this simple example. There are, however, certain additional instructions which are required for a working version of the program to run on the computer and those are included in the complete listing below.

```
*      CSMP PROGRAM FOR RELATIVE GROWTH MODEL

*          PROBLEM DEFINITION
*            --STARTING WEIGHT=20.0
*            --RELATIVE GROWTH RATE=0.01 PER DAY
*            --CALCULATE GROWTH OVER 50 DAYS
*          INITIALISATION AND OTHER DATA INPUTS
PARAMETER IWT=20.0,RGR=0.01                             001
*          FLOWS
        AGR=WT*RGR                                      002
*          COMPARTMENTS
        WT=INTGRL(IWT,AGR)                              003
*          ADDITIONAL INSTRUCTIONS
*            --INTEGRATION METHOD
METHOD RECT                                             004
*            --TIME STEP,TOTAL TIME AND PRINTING
*              AND PLOTTING INTERVALS
TIMER DELT=1.0,FINTIM=50.0,PRDEL=1.0,OUTDEL=1.0         005
*            --HEADINGS FOR PRINTED AND PLOTTED
*              OUTPUT
TITLE RELATIVE GROWTH MODEL                             006
LABEL RELATIVE GROWTH MODEL                             007
*            --VARIABLES TO BE PRINTED AND PLOTTED
PRINT AGR,WT                                            008
OUTPUT WT                                               009
END                                                     010
STOP                                                    011
```

Lines in the complete listing of the program which begin
with an asterisk are for reference purposes only and are
ignored by the computer. Of the remaining lines, constituting the instructions to the computer and referenced for
our purposes by serial numbers in the right-hand margin,[1]
we have dealt with lines 1 to 4 above.

Line 5 contains instructions on the time-step between
successive calculations, DELT, on the total time period,
FINTIM and on the intervals at which the results are to be
printed and plotted, PRDEL and OUTDEL. Notice that all
these time periods are shown as numerical values only, with
no explicit reference to the units involved. This reflects
the fact that in such programs the user must decide, in advance, on common units for measuring time and other variables
in the program as a whole and construct the program so that
the units for different variables are either directly com-
patible or are converted as appropriate. In this case time
is given in days and weight as kilograms.

Headings for the printed and plotted output are
specified in lines 6 and 7. Line 8 specifies the variables
for which numerical values are to be printed out and line
9 is an instruction to plot out the named variables in
graphical form. In both these forms of output from the
program, time within the simulation period is provided
automatically in the results, without being specifically
requested by the user. The results for this example as
produced by the computer are shown in Figs. 3.2 and 3.3.

---

[1] In a working version of the program to be presented to the computer it
is necessary to place any such reference numbers in columns 73 through
80 of the standard, 80-column lines which are punched on computer cards
or tape. Additional conventions concerning layout within each line
of a CSMP program are that normal structure statements commence in
column 7, but data and control statements like PARAMETER and TIMER
begin in column 1. Where a statement cannot be accommodated within a
single line up to column 72 it may be continued, with certain restric-
tions, on one or more additional lines, provided that the user indicates
a 'continuation card' is to follow by ending the previous line with
three consecutive decimal points, thus ... . The reader should consult
the users' manual for further details of CSMP usage in program layout
(IBM 1975).

RELATIVE GROWTH MODEL

TIME	AGR	WT
.0	.20000	20.000
1.00000	.20200	20.200
2.00000	.20402	20.402
3.00000	.20606	20.606
4.00000	.20812	20.812
5.00000	.21020	21.020
6.00000	.21230	21.230
7.00000	.21443	21.443
8.00000	.21657	21.657
9.00000	.21874	21.874
10.0000	.22092	22.092
11.0000	.22313	22.313
12.0000	.22536	22.536
13.0000	.22762	22.762
14.0000	.22989	22.989
15.0000	.23219	23.219
16.0000	.23451	23.451
17.0000	.23686	23.686
18.0000	.23923	23.923
19.0000	.24162	24.162
20.0000	.24404	24.404
21.0000	.24648	24.648
22.0000	.24894	24.894
23.0000	.25143	25.143
24.0000	.25395	25.395
25.0000	.25648	25.648
26.0000	.25905	25.905
27.0000	.26164	26.164
28.0000	.26426	26.426
29.0000	.26690	26.690
30.0000	.26957	26.957
31.0000	.27226	27.226
32.0000	.27499	27.499
33.0000	.27774	27.774
34.0000	.28051	28.051
35.0000	.28332	28.332
36.0000	.28615	28.615
37.0000	.28901	28.901
38.0000	.29190	29.190
39.0000	.29482	29.482
40.0000	.29777	29.777
41.0000	.30075	30.075
42.0000	.30375	30.375
43.0000	.30679	30.679
44.0000	.30986	30.986
45.0000	.31296	31.296
46.0000	.31609	31.609
47.0000	.31925	31.925
48.0000	.32244	32.244
49.0000	.32567	32.567
50.0000	.32892	32.892

Fig.3.2. Printed output from the relative growth model.

Readers familiar with the more detailed and explicit instructions required to produce similar results from running
a program using FORTRAN will appreciate the simplicity with
which this task may be dealt with in CSMP. Reference should
be made to the users' manual for details of the wide range of
simplified output instructions which is available in that
language.

RELATIVE GROWTH MODEL

'+'=WT

19.50   34.50

TIME	WT
.0	20.000
1.0000	20.200
2.0000	20.402
3.0000	20.606
4.0000	20.812
5.0000	21.020
6.0000	21.230
7.0000	21.443
8.0000	21.657
9.0000	21.874
10.000	22.092
11.000	22.313
12.000	22.536
13.000	22.762
14.000	22.989
15.000	23.219
16.000	23.451
17.000	23.686
18.000	23.923
19.000	24.162
20.000	24.404
21.000	24.648
22.000	24.894
23.000	25.143
24.000	25.395
25.000	25.648
26.000	25.905
27.000	26.164
28.000	26.426
29.000	26.690
30.000	26.957
31.000	27.226
32.000	27.499
33.000	27.774
34.000	28.051
35.000	28.332
36.000	28.615
37.000	28.901
38.000	29.190
39.000	29.482
40.000	29.777
41.000	30.075
42.000	30.375
43.000	30.679
44.000	30.986
45.000	31.296
46.000	31.609
47.000	31.925
48.000	32.244
49.000	32.567
50.000	32.892

Fig.3.3. Plotted output from the relative growth model.

In lines 10 and 11 the END instruction indicates that an individual run of the model has been completely specified and the STOP command similarly signifies the completion of any 're-runs', with the same model structure but with any required alterations to parameter values. In this case, no instructions to change numerical values are inserted between the two commands as re-runs are not required.

3.3.2. The essentials of a FORTRAN program

We give below the core part of a FORTRAN program to perform the operations for the relative growth model which are effected by lines 1-4 in the CSMP program. Detailed instructions for a complete, working program in FORTRAN are included in Appendix II, with specially designed subroutines to allow the user to produce similar output of results to that from the CSMP version with minimum attention to the technical details.

```
C             CORE PART OF A FORTRAN PROGRAM
C             FOR THE RELATIVE GROWTH MODEL
C          INITIALISATION AND DATA INPUTS
           WT=20.0                                    1
           RGR=0.01                                   2
C          TIME LOOP SET TO NUMBER OF
C          CALCULATION INTERVALS IN TOTAL
C          RUN TIME
           DO 100 ITIME=1,50                          3
C          FLOWS
           AGR=WT*RGR                                 4
C          COMPARTMENTS
           WT=WT+AGR                                  5
C          END OF TIME LOOP
       100 CONTINUE                                   6
```

The convention used for indicating users comments, to be ignored by the computer, is to commence such lines with a 'C'.

Again, the remaining lines are serially numbered in the right-hand margin for reference purposes.

Lines 1 and 2 are used to set the initial condition for the compartment and to specify the relative growth rate. Next, an iterative mechanism in the form of a 'DO loop' is set up to repeat the calculations for the rate of growth and updating the compartment, according to how many time steps are involved in the total run-time for the model. In this instance, fifty iterations are specified and the instruction in line 3 allows for the counter, ITIME, to be incremented in steps of unity between 1 and 50. The extent of the loop, defining which instructions are to be repeated, is specified by the 'scope label' of '100' and which appears both in line 3 and in the CONTINUE' instruction in line 6, i.e. the calculations to be repeated appear between lines 3 and 6. Those calculations will be performed each time the counter is incremented until it reaches its maximum value of 50. Thereafter, any remaining instructions, beyond line 6, will be executed.

The *order* of instructions in the FORTRAN version is an integral and important part of the overall program. The user may rely therefore on the overall sequence being as follows:

(1)   The lines preceding the DO loop are executed once only and in the order in which they are written.
(2)   The instructions within the DO loop are repeated the specified number of times, again in the exact order in which they are set out.
(3)   Any instructions following the DO loop are carried out once only and in the given order.

The effect is that the initial value for WT which is given in line 1 is read in at the beginning of the operations and that value is the one used for calculating the rate of growth in line 4 during the *first* passage through the DO loop. The next calculation during the first passage through the loop is to take the value for the growth rate, AGR, which has just been calculated in line 3 and to use it in line 4 to add on to the initial value for WT and thus arrive at the updated

value for the compartment. During the second repetition
of the calculations in the loop, the updated value for WT
produced the first time around is the one used in line 3 for
calculating the next growth increment, and so on.

Two general rules which apply to the sequence of calculations in a FORTRAN program for this type of model are:

(1)   data inputs, including initial conditions, precede
       the time loop,
(2)   within the time loop, flow calculations precede up-
       dating operations for the compartments.

One consequence of the strictly sequential form in which
the FORTRAN program is written is that no distinct names
are required for the initial conditions of the compartments.
This is in contrast to the CSMP version, where IWT refers
specifically to the starting weight. This relatively minor
distinction in the outward form of the programs reflects a fundamental difference in the way they are handled in the computer. There is no requirement to set down the individual
steps in the calculations in the correct calculation sequence
in CSMP because in the first of two stages of 'compiling'
the program, prior to carrying out the calculations, a built-
in sorting algorithm rearranges the order of the statements
as necessary. For example, lines 2 and 3 in the CSMP program
could be presented in the reverse order to that shown and
this would not affect the correct interpretation and running
of the program.[1]

Mention may be made here also of the facility in CSMP
for dividing up a program into what are called *Initial*,
Dynamic, and *Terminal* sections. Using this optional facility
to separate *Initial* and *Dynamic* sections is equivalent to the
division of a FORTRAN program into the sections preceding

---

[1] It is permissible, also, to include within a CSMP program one or more
specially labelled sections of FORTRAN code which are not subject to
the sorting algorithm. This facility is of value when the user wishes
to impose a particular sequence of operations. Again, reference should
be made to the users' manual for the details of program construction
involved.

and within the time loop. Thus the *Initial* section contains statements concerned with the initial values of compartments and other numerical values which do not vary throughout the execution of the model. Within the *Dynamic* section are set out the instructions on model structure and its implementation in terms of rate calculations and updating the compartments. Separation of *Initial* and *Dynamic* sections is economical of computer time in that it ensures that the instructions in the *Initial* part are executed once only and not repeated, as they would be without the division, every time the sequence of calculations for updating the compartments is carried out. Such economy is of little significance in short programs such as that for the relative growth model, but can be important in large models. The use of a *Terminal* section is designed chiefly to facilitate carrying out calculations between successive runs of a model where the values of reset parameters are to be derived from evaluating the model results in the previous run.

3.4. APPROXIMATIONS IN NUMERICAL INTEGRATION, WITH SPECIAL REFERENCE TO THE RELATIVE GROWTH MODEL

We have noted that numerical integration methods are necessarily approximate and whilst it would be inappropriate in this context to attempt a general exposition of the nature and extent of those approximations, it is important to be aware of some of the basic principles involved and how these relate to the practice of modelling agricultural systems. The relative growth model is a convenient example for this purpose as it is sufficiently simple to allow an exact, analytical solution with which the results of numerical methods may be compared.

In the relative growth model the rate of change in weight of the organism may be expressed in mathematical terms in the following equation

$$\frac{dW}{dt} = RGR \times W \quad ,$$

where dW and dt are infinitesimally small increments of weight and time, RGR the relative growth rate and W the weight of the organism.

Integration of that equation gives

$$W = IW \times \exp(RGR \times T),$$

where IW is the initial weight of the organism.

If we choose a set of numerical values for the terms on the right-hand side of the equation we can solve for W, e.g. if $IW = 1.0$, $RGR = 0.2$, $T = 5.0$, then

$$W = 1.0 \times \exp(0.2 \times 5.0)$$

$$= e$$

$$= 2.71828 \ldots .$$

With this particular set of numerical values, W is equal to the well-known mathematical constant e, so we can use this as a yardstick for looking at approximate solutions produced by numerical methods.

As a start, let us examine a numerical solution, using rectangular integration and advancing time in units of one (Table 3.1). Comparing the exact solution of 2.71828 with

TABLE 3.1

Time	Weight	Weight increment
0	1.0	0.2
1	1.2	0.24
2	1.44	0.288
3	1.728	0.3456
4	2.0736	0.41472
5	2.48832	

the approximation of 2.48832, there is a noticeable under-estimation. Evidently, such a coarse subdivision of the total

time period produces a relatively inexact estimate when contrasted with the analytical solution, in which the relative
growth rate is applied to infinitesimally small periods of
time. As an exercise, the reader may repeat the calculations
using a time-step of half a unit and verify that the closer
approximation of 2.59374 results from using that smaller
interval. Note that once we depart from a time-step of unity
it is necessary to account for this in the calculation of each
weight increment and the expression that must be used is

$$INC = RGR \times W \times DT$$

where DT is the time-step or 'delta-time'.

The calculations become successively more tedious to
accomplish by hand with smaller time-steps and the reader may
examine the effects for himself by inserting the appropriate
numerical values in the computer program given in Section 3.3.
But to illustrate the overall effects of reducing the time-
step on solution accuracy, the results for a limited range of
calculations are shown in Table 3.2.

TABLE 3.2

Time-step	Solution value
1.0	2.4883
0.5	2.5937
0.1	2.6916
0.01	2.7152
0.001	2.7161
.	.
.	.
.	.
Analytical solution	2.7183

Clearly, the numerical solutions approach the analytical
one more closely as the time-step is successively reduced;

but an important feature is that the successive gains in
accuracy become smaller for equivalent reductions in the
length of the time-step as the numerical solutions converge on
the analytical solution. With a decreasing return in terms
of solution accuracy for the investment of additional computer
time to perform the extra calculations there is a practical
incentive to effect a compromise, at some point, between
accuracy and the expense involved.

It must be emphasized that in the above comparisons
between the numerical and analytical solutions we are assuming
that the given value for the relative growth rate is the true,
'instantaneous' one. If, as is frequently the case in practice, the only estimate of relative growth rate available is
derived from successive measurements of weight or other attributes of growth which are recorded at finite time-intervals
then it is invalid to treat such an estimate as a true reflection of the instantaneous value. In those circumstances
the accuracy of the model solution is constrained by the data
available and there is no justification for integrating over
much smaller time-steps than those of the original measurements.[1]

Neither is it usual in practice, of course, to have an
analytical solution available for comparison with those from
numerical methods, as is the case with this simple example.
Indeed, the principal reason for employing numerical methods
is the difficulty or impossibility of devising analytical
solutions for the majority of applications. In such situations, it can be helpful to experiment with varying time-
steps in an empirical fashion. de Wit and Goudriaan (1974)
suggest the simple approach of assuming that an appropriate
time-step 'is reached when halving its value does not change
the relevant results of the simulation by more than a pre-
set, relative amount'.

It should be appreciated, of course, that such tests are
empirical and not foolproof if used in a purely mechanical

[1] See p.84 for further discussion of appropriate calculation intervals
in relation to the representation of feedback loops.

fashion. It is particularly important not to lose sight of the background in which a model is being built, in terms of the sources and reliability of the data used. We refer later (p.91) to some special applications where accuracy of numerical solutions can be an important consideration; but in the majority of biological and agricultural models the 'data-base' is so approximate that it is unrealistic to do more than avoid gross errors in the integration process.

For the same reason, it is rarely appropriate to consider the use of more sophisticated integration methods which can increase solution accuracy. But for the sake of completeness we mention the use of *variable length time-steps*, employing short steps when the rate processes are changing rapidly and longer ones when the changes are slower. Some algorithms allow the user to combine economy of computer usage with a desired level of solution accuracy by permitting the specification of maximum error bounds for each integration step, either absolute or relative, when using variable step-lengths. CSMP provides a wide range of integration methods and the interested reader should consult the users' manual for a general discussion of their applications. Examples of particular biological applications are given in the monographs by de Wit and Van Keulen (1972), de Wit and Goudriaan (1974), and Frissel and Reiniger (1974).

3.5. PROGRAMMING THE MODEL OF CARBON METABOLISM IN A GREEN PLANT

This second example of programming a dynamic simulation model is based on the flow diagram, Fig.2.3 on page 26 .

The following numerical inputs are used in the program:

Initial weight of photosynthetic organs = 300 mg
Initial weight of non-photosynthetic organs = 200 mg
Initial content of carbohydrate pool = 0.0
Proportion of carbohydrate translocated to photosynthetic organs = 70 per cent
Proportion of carbohydrate translocated to non-photosynthetic organs = 30 per cent.

The initial rate of photosynthesis is 0.1 mg per mg
of photosynthetic tissue per day; this unit rate
declines linearly to 0.075 mg on day 10, to 0.05
on day 20, and 0.04 on day 90.

The respiration rate of photosynthetic tissue is
taken as 0.015 mg per mg of tissue per day and
the corresponding rate for non-photosynthetic
tissue is 0.02.

Simulation period = 90 days.

3.5.1. A CSMP program

A complete program in CSMP is given below, using the same
layout as that for the relative growth model.

```
*      CSMP PROGRAM FOR MODEL OF CARBON METABOLISM
*      IN A GREEN PLANT

*      INITIALISATION AND OTHER DATA INPUTS
INITIAL                                                            001
PARAMETER ICPO=300.0,ICNPO=200.0,PRT1=0.7,...                      002
          PRT2=0.3,PRR1=0.015,PRR2=0.02                            003

DYNAMIC                                                            004

*         FLOWS
       RPHOT=PO*URATE                                              005
       URATE=AFGEN(TRATE,TIME)                                     006
FUNCTION TRATE=(0.0,0.10),(10.0,0.075),(20.0,0.05),...            007
          (90.0,0.04)                                              008

       RT1=CHP*PRT1                                                009
       RT2=CHP*PRT2                                                010

       RR1=PO*PRR1                                                 011
       RR2=NPO*PRR2                                                012

*         COMPARTMENTS
       CHP=INTGRL(0.0,(RPHOT-RT1-RT2))                             013
       PO=INTGRL(ICPO,(RT1-RR1))                                   014
       NPO=INTGRL(ICNPO,(RT2-RR2))                                 015

*         ADDITIONAL INSTRUCTIONS
METHOD RECT                                                        016
TIMER DELT=1.0,FINTIM=90.0,PRDEL=1.0,OUTDEL=2.0                    017
TITLE      MODEL OF CARBON METABOLISM                              018
LABEL      MODEL OF CARBON METABOLISM                              019
PRINT PO,NPO                                                       020
OUTPUT PO,NPO                                                      021
PAGE GROUP,SYMBOL=(P,N),NTAB=0                                     022
END                                                                023
PARAMETER PRR1=0.005,PRR2=0.01                                     024
END                                                                025
STOP                                                               026
```

As in the relative growth model, the first step in the program
is to supply the required data on initial conditions and on
the values of constants, in lines 2 and 3. Notice that in
this instance the program has been formally divided into
Initial and *Dynamic* sections by the instructions in lines 1
and 4. Within the parameter list, we set the initial weights
for the photosynthetic and non-photosynthetic organs (ICPO

and ICNPO); those values are followed by the proportions of
material translocated from the carbohydrate pool (PRT1 and
PRT2) and, finally, by the respiration constants (PRR1 and
PRR2).

The first set of instructions in the *Dynamic* section sets
out the calculations for the entry of carbon into the system,
the rate of photosynthesis. In line 5 this flow is specified
as the product of the amount of photosynthetic tissue (PO)
and the rate per unit of that tissue (URATE). Lines 6-8
define the unit rate as an arbitrary function of time, using
the special CSMP function called AFGEN. We first give notice,
in line 6, that we are defining such a function, that its
name is TRATE and that TRATE is to vary according to the
value of TIME. The CSMP compiler recognizes the variable
TIME without its being specially defined by the user because
it is automatically created for use as the basic independent
variable in a model run. The pairs of numerical values en-
closed by parentheses in lines 7 and 8 give, in each case, the
value of the independent variable, TIME, followed by that
for TRATE. Thus, at time zero TRATE has a value of 0.1,
decreasing to 0.075 on the tenth day, and so on. Given such
pairs of values, the AFGEN function will generate any intermediate values that are required to run the model by linear
interpolation, as illustrated in Fig.3.4. If it should happen

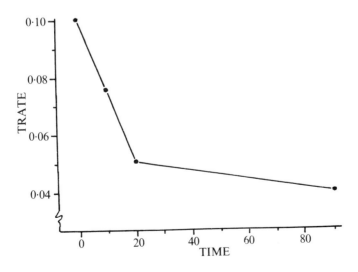

Fig.3.4. Graph of the relationship between TRATE and TIME showing the
effect of linear interpolation with the AFGEN function.

that values are required which lie outside the range of the quoted figures then the extreme value from the appropriate end of the range will be used; but the user will be warned that this step has been necessary by an 'error message' printed with the model results.

Use of the AFGEN function is frequently the most convenient way of specifying a relationship when the only information available is from a small number of experimental observations which do not justify an elaborate curve-fitting procedure. If the relationship can be defined by an algebraic equation this is a more economical and satisfactory procedure.

Lines 9 and 10 in the program specify the rates of translocation from the carbohydrate pool to the photosynthetic and non-photosynthetic organs, using the proportionality constants PRT1 and PRT2. Note that the simplifying assumptions underlying this particular model include the complete allocation of all material in the pool and a 1:1 conversion of carbohydrate into plant tissue.

The rates of respiration in lines 11 and 12 again use proportionality constants (PRR1 and PRR2) to define their dependence on the weights of tissue present.

The three compartments or state variables are specified in lines 13 to 15, using INTGRL statements. Note that where more than one rate process is involved in the definition of a compartment an additional set of parentheses encloses the set of rate variable abbreviations. The initial conditions for the compartments representing the photosynthetic and non-photosynthetic organs are specified, as in the relative growth model, by using separate variable names which are assigned numerical values in the PARAMETER statement. For the carbohydrate pool we use the alternative convention of inserting the appropriate numerical value directly in the INTGRL statement.

In line 22 of this example we use an additional instruction to the standard OUTPUT command in order to control the form of the graphical output. A number of features may be modified by this PAGE statement (see CSMP users' manual). Here, it is used to ensure that all the variables are plotted

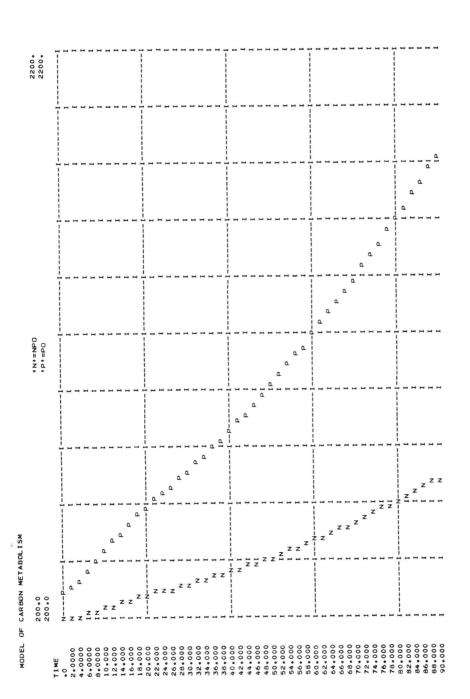

Fig.3.5. Plotted output from the carbon metabolism model, with PRRI = 0.015 and PRR2 = 0.02.

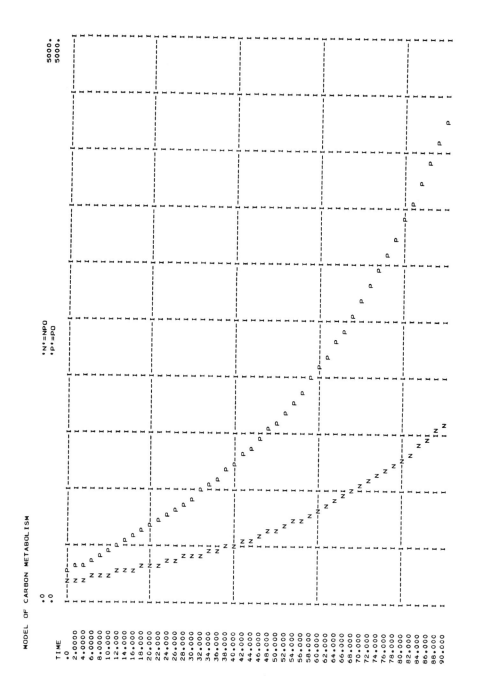

Fig.3.6. Plotted output from the carbon metabolism model with PRRI = 0.005 and PRR2 = 0.01.

on a common scale (GROUP), to assign chosen symbols to the
variables (SYMBOL), and to suppress the printing of numerical
values for the variables (NTAB).

The remainder of the program follows the format used
in the relative growth model except that an additional run is
specified by alternative values for PRR1 and PRR2 between the
END and STOP commands. Notice the additional END command which
follows each revision of parameter values for an extra run of
the model. Examples of the plotted output from this example are
given in Figs.3.5 and 3.6.

3.5.2. The essentials of a FORTRAN program
As in the other examples, we give here only the part of the
FORTRAN program which performs the basic calculations to de-
fine the rate processes and up-date the state variables.
Details of the complete program are given in Appendix II.

```
C               CORE PART OF A FORTRAN PROGRAM
C               FOR THE MODEL OF CARBON
C               METABOLISM IN A GREEN PLANT
C          INITIALISATION AND DATA INPUTS
           CHP=0.0                                  1
           PO=300.0                                 2
           NPO=200.0                                3
           PRT1=0.7                                 4
           PRT2=0.3                                 5
           PRT1=0.015                               6
           PRR2=0.020                               7
C          TIME LOOP
           DO 300 ITIME=1,90                        8
           FLOWS
           URATE=TABFN(TRATE,4,ITIME)               9
           RPHOT=PO*URATE                          10
           RT1=CHP*PRT1                            11
           RT2=CHP*PRT2                            12
           RR1=PO*PRR1                             13
           RR2=NPO*PRR2                            14
C          COMPARTMENTS
           CHP=CHP+RPHOT-RT1-RT2                   15
           PO=PO+RT1-RR1                           16
           NPO=NPO+RT2-RR2                         17
C          END OF TIME LOOP
       300 CONTINUE                                18
```

Lines 1 to 7 supply the numerical values for initial conditions and constants prior to entering the time loop. Line
8 sets up the time loop, which extends as far as line 18 and
contains the calculations of rate processes and updating of
the compartments.

Of the calculations for the transfer rates, that for the
entry of carbon in the process of photosynthesis is the only
one which differs from the CSMP version. In calculating photosynthesis, use is made of a specially designed subroutine which
is essentially equivalent to the AFGEN function in CSMP and is
described in Appendix II. This equivalent to the AFGEN function, designated TABFN for 'table function', requires the
user to specify not only the names of the dependent and independent variables but also the number of pairs of numerical
values which are involved, four in this case and entered as
the second term within the parentheses. Notice, also, the
order of the statements, with the definition of URATE preceding that for RPHOT. In the CSMP version the equivalents
of these two statements appear in the reverse order, a point
which is not critical since the sorting algorithm will re-
arrange them in the correct sequence for computation, as required.

The equivalents of the INTGRL statements in the CSMP
version appear in lines 15 to 17 of the FORTRAN program as
instructions to perform the necessary additions and subtractions. As explained in relation to the program for the relative growth model, the *order* of instructions in the FORTRAN
program ensures that the appropriate numerical values for the
initial conditions of the compartments are used in the first
passage through the time loop. Thereafter, the numerical
values for the contents of the compartments which were calculated in the previous iteration are the starting points for
the next round of up-dating in lines 15 - 17. The strict
sequence of operations ensures, also, that in each passage
through the loop the rates are calculated first and these
results are available for up-dating the compartments.

3.6. PROGRAMMING THE ANIMAL POPULATION MODEL
This third example of programming a simulation model is based

on the flow diagram, Fig.2.4. (p. 29).

In programming the model we assume that we are concerned with an animal population in which young are borne annually and that both males and females take exactly one year to reach maturity. At the start of the simulation period of 50 years the following numbers of animals are present in the various classes

Young females	70
Young males	60
Adult females	200
Adult males	55

The mean overall birth rate is taken as 60 per hundred adult females, with a random variation in individual years such that, on average, half the actual figures lie between the 55 and 65 per cent points, a quarter within the range 45 to 55 per cent and a further quarter between 65 and 75 per cent.

The sex ratio is assumed to be invariant, in the proportion 48 males to 52 females.

Mean death rates between birth and maturity are 5 per cent for females and 7 per cent for males. Random variation of the death rates in individual years is such that the average proportions shown in Table 3.3 occur.

TABLE 3.3

Females		Males	
Death rate (%)	Proportion	Death rate (%)	Proportion
0-4	0.25	2-6	0.25
4-6	0.50	6-8	0.50
6-10	0.25	8-12	0.25

It is assumed that there are no deaths of adult animals. Twenty-nine per cent of the adult females are sold each year and all the adult males are sold, ignoring the small number of males that would be retained, in practice, for breeding purposes.

3.6.1. A CSMP program

A program for this model in CSMP is set out below, using the same overall layout as in the previous examples.

```
*       CSMP PROGRAM FOR ANIMAL POPULATION MODEL


*           INITIALISATION AND OTHER DATA INPUTS
INITIAL                                                             001
PARAMETER ICYF=70.,ICAF=200.,ICYM=60.,ICAM=55.,...                  002
          PRNF=0.52,PRNM=0.48,PSLF=0.29,PSLM=1.0                    003

DYNAMIC                                                             004
*       FLOWS

*           BIRTHS
        OBR=AF*PBR                                                  005
        PBR=AFGEN(TPBR,RNUM1)                                       006
FUNCTION TPBR=(0.0,0.45),(0.25,0.55),(0.75,0.65),...               007
             (1.0,0.75)                                             008
        RNUM1=RNDGEN(5)                                             009
        BRF=OBR*PRNF                                                010

        BRM=OBR*PRNM                                                011

*           DEATHS
        DRF=YF*PDRF                                                 012
        PDRF=AFGEN(TPDRF,RNUM2)                                     013
FUNCTION TPDRF=(0.0,0.0),(0.25,0.04),(0.75,0.06),...               014
             (1.0,0.1)                                              015
        RNUM2=RNDGEN(3)                                             016

        DRM=YM*PDRM                                                 017
        PDRM=AFGEN(TPDRM,RNUM3)                                     018
FUNCTION TPDRM=(0.0,0.02),(0.25,0.06),(0.75,0.08),...              019
             (1.0,0.12)                                             020
        RNUM3=RNDGEN(9)                                             021

*           SURVIVAL TO MATURITY
        SRF=YF*(1.0-PDRF)                                           022
        SRM=YM*(1.0-PDRM)                                           023

*           SALES
        SLRF=AF*PSLF                                                024
        SLRM=AM*PSLM                                                025

*           COMPARTMENTS
        YF=INTGRL(ICYF,(BRF-DRF-SRF))                               026
        YM=INTGRL(ICYM,(BRM-DRM-SRM))                               027
        AF=INTGRL(ICAF,(SRF-SLRF))                                  028
        AM=INTGRL(ICAM,(SRM-SLRM))

        TOTSF=INTGRL(0.0,SLRF)                                      030
        TOTSM=INTGRL(0.0,SLRM)                                      031

*           ADDITIONAL INSTRUCTIONS
METHOD RECT                                                         032
TIMER DELT=1.0,FINTIM=50.0,PRDEL=1.0                                033
TITLE    ANIMAL POPULATION MODEL                                    034
PRINT YF,AF,TOTSF,YM,AM,TOTSM                                       035
END                                                                036
STOP                                                                037
```

Line 2 of the INITIAL section sets the starting values for the four compartments, supplying data on the numbers of young males and females (ICYF and ICYM) and also on the numbers in the adult classes (ICAF and ICAM). In line 3, the proportions of female (PRNF) and male (PRNM) births are specified and the selling rates for females (PSLF) and males (PSLM) are given as proportions of the numbers of adult animals.

The overall birth rate is defined, in line 5, as the product of the number of adult females and the proportionality factor, PBR. Next, PBR is set equal to an arbitrary function generator in which the value returned by the function depends on a random number, RNUM1. The independent variable, RNUM1, is derived from a 'generator', in line 9, which produces random numbers with an even distribution between zero and one. The pairs of values for the dependent and independent variables within the AFGEN function are so chosen that random numbers between values of 0.25 and 0.75 are linked to values of PBR in the range 0.55 to 0.65. The effect of that link is that for one-half of the total range of RNUM1, PBR will take on values between the 55 and 65 per cent points. Similarly, for one-quarter of the total range of RNUM1, PBR will lie between the 45 and 55 per cent points and a further quarter of the range will result in values for PBR between the 65 and 75 per cent points.

It should be noted that this treatment of the overall birth rate as a stochastic process and the similar mechanisms to simulate the death rates described below are designed to illustrate only the general form that may be employed. To ensure valid results from a model with stochastic elements it is necessary to take account of a number of technical points. For example, the numbers which are produced from so-called random-number generators like the one used here are not genuinely random, but 'pseudo-random'. The choice of a particular *seed* number, 5 in this instance, uniquely determines a particular sequence of numbers which will result from successive calls on the generator and the user must be aware of that feature in designing a particular application. Another important consideration is that the model should be set up and run in such a way that the results constitute an appropriate sample of the overall behaviour of the system for statistical analysis. The reader is referred to Davies (1971) for a more detailed discussion of the generation of random numbers. Naylor, Balintfy, Burdick, and Chu (1966) provides a comprehensive treatment of stochastic modelling.

Having accounted for the variation in the overall birth rate, it remains to apportion the births to males and females in lines 10 and 11.

Death rates of young males and females are calculated in lines 12 to 21, using exactly analogous mechanisms to that for the overall birth rate to mimic their stochastic behaviour. Notice that in each case the order of the individual statements in the program does not correspond to the sequence in which the calculations need to be performed. Advantage is taken of the CSMP sorting algorithm to set down first the basic rate equation and then to consider, successively, what calculations are needed to support that rate equation. For example, in line 12 we define the rate of death of young females as the product of the number of animals in that class and a proportionality factor, PRDF. Then in lines 13-15 we define PRDF as the output from an arbitrary function generator. In turn, the arbitrary function generator requires a sequence of random numbers as its independent variable and that need is met by the generator in line 16. Whether such a sequence of reasoning appears convenient or 'logical' is largely a matter of individual taste, of course.

With the simplifying assumption that both males and females take exactly one year to reach maturity and may be sold or used for breeding at that point it is possible to settle for a calculation interval of one year in the model. As a consequence, the equations for the survival rates of both young females and males in lines 22 and 23 are concerned only with the simple operations of transferring the contents of the young-animal compartments to the adult-animal compartments, with appropriate deductions for the death-rates over a year.

Sales rates are defined in lines 24 and 25 as the products of the pre-determined proportions of animals to be sold and the numbers of animals in the adult classes.

ANIMAL POPULATION MODEL

TIME	YF	AF	TOTSF	YM	AM	TOTSM
.0	70.000	200.00	.0	60.000	55.000	.0
1.00000	46.814	212.00	58.000	43.213	58.794	55.000
2.00000	49.696	197.32	119.48	45.873	42.325	113.79
3.00000	46.545	189.75	176.70	42.965	44.841	156.12
4.00000	45.832	181.10	231.73	42.307	41.675	200.96
5.00000	47.498	173.78	284.25	43.844	39.871	242.64
6.00000	54.027	168.86	334.64	49.871	39.667	282.51
7.00000	56.426	173.14	383.61	52.086	44.215	322.17
8.00000	62.963	176.00	433.82	58.119	49.837	366.39
9.00000	62.140	184.49	484.86	57.360	54.736	416.22
10.0000	46.041	192.65	538.36	42.499	55.027	470.96
11.0000	46.957	180.01	594.23	43.345	39.310	525.99
12.0000	58.177	173.91	646.43	53.702	40.146	565.30
13.0000	48.795	177.05	696.87	45.042	47.839	605.44
14.0000	64.463	172.99	748.21	59.505	42.415	653.28
15.0000	49.023	184.49	798.38	45.252	57.871	695.70
16.0000	58.429	178.04	851.88	53.934	43.453	753.57
17.0000	46.623	181.82	903.51	43.037	50.542	797.02
18.0000	67.020	171.54	956.24	61.865	39.096	847.56
19.0000	46.451	184.80	1006.0	42.878	60.304	886.66
20.0000	71.566	177.04	1059.6	66.061	39.136	946.96
21.0000	52.099	192.91	1110.9	48.092	61.494	986.10
22.0000	48.978	185.62	1166.9	45.210	44.320	1047.6
23.0000	59.994	176.40	1220.7	55.379	40.084	1091.9
24.0000	60.937	182.24	1271.9	56.250	51.819	1132.0
25.0000	51.944	187.01	1324.7	47.948	52.279	1183.8
26.0000	56.404	183.36	1378.9	52.065	43.170	1236.1
27.0000	50.248	183.80	1432.1	46.383	48.202	1279.3
28.0000	58.384	178.35	1485.4	53.893	42.482	1327.5
29.0000	57.378	182.59	1537.1	52.964	51.032	1369.9
30.0000	58.332	183.63	1590.1	53.845	49.764	1421.0
31.0000	46.408	184.88	1643.3	42.838	49.643	1470.7
32.0000	55.645	176.49	1697.0	51.364	38.641	1520.4
33.0000	56.042	176.30	1748.1	51.731	46.312	1559.0
34.0000	60.471	176.08	1799.3	55.820	48.259	1605.3
35.0000	57.960	182.21	1850.3	53.502	51.041	1653.6
36.0000	43.535	186.35	1903.2	40.186	48.288	1704.6
37.0000	49.683	174.51	1957.2	45.861	38.810	1752.9
38.0000	58.582	172.08	2007.8	54.075	41.808	1791.7
39.0000	65.507	178.00	2057.7	60.468	52.876	1833.5
40.0000	48.385	186.98	2109.3	44.663	54.570	1886.4
41.0000	58.224	179.41	2163.6	53.745	43.517	1941.0
42.0000	52.770	182.07	2215.6	48.710	50.035	1984.5
43.0000	57.216	179.37	2268.4	52.815	45.228	2034.5
44.0000	48.560	182.18	2320.4	44.825	46.973	2079.8
45.0000	54.272	176.81	2373.2	50.097	40.799	2126.7
46.0000	56.963	178.09	2424.5	52.581	46.799	2167.5
47.0000	51.978	178.42	2476.2	47.980	50.900	2214.3
48.0000	54.908	175.54	2527.9	50.684	43.112	2265.2
49.0000	68.352	177.96	2578.8	63.094	46.871	2308.4
50.0000	63.832	191.67	2630.4	58.922	57.230	2355.2

Fig.3.7. Printed output from the animal population model.

The reader may check, by reference to the flow diagram
(Fig.2.4, p.29), that the integral equations for the compartments are constructed by considering, in each case, which
rates add to and which subtract from the contents of a given
compartment. Note the use of the alternative format of inserting a numerical value for the initial condition directly
in the INTGRL equation itself in lines 30 and 31. Those compartments, for the totals of male and female animals sold
during a run of the model, are 'dummy' compartments used

simply to accumulate the totals and that purpose requires that the initial contents be zero, of course.

The remainder of the program follows the pattern of previous examples except that in this instance we request only the printed form of output (Fig.3.7), omitting the OUTDEL specification from the TIMER statement and the corresponding OUTPUT command.

3.6.2. *The essentials of a FORTRAN program*

As in the previous examples, we give here only the part of a FORTRAN program which is concerned with the definitions of the rate processes and the updating of the compartments. A complete listing of the FORTRAN version is included in Appendix II.

```
C            CORE PART OF A FORTRAN PROGRAM
C            FOR THE ANIMAL POPULATION MODEL
C         INITIALISATION AND DATA INPUTS
      DATA YF, AF, YM, AM, TOTSF, TOTSM/
      A70.,200.,60.,55.,0.,0./                      1
      DATA PRNF,PRNM,PSLF,PSLM/0.52,0.48,0.29,1./   2
C         TIME LOOP
      DO 100 ITIME=1,50                             3
C         FLOWS
C            BIRTHS
      RNUM1=RNDX(IX)                                4
      PBR=TABFN(TPBR,4,RNUM1)                       5
      OBR=AF*PBR                                    6
      BRF=OBR*PRNF                                  7
      BRM=OBR*PRNM                                  8
*            DEATHS
      RNUM2=RNDX(IX)                                9
      PDRF=TABFN(TPDRF,4,RNUM2)                     10
      DRF=YF*PDRF                                   11
      RNUM3=RNDX(IX)                                12
      PDRM=TABFN(TPDRM,4,RNUM3)                     13
      DRM=YM*PDRM                                   14
C            SURVIVAL TO MATURITY
      SRF=YF*1.0-PDRF                               15
      SRM=YM*1.0-PDRM                               16
C            SALES
      SLRF=AF*PSLF                                  17
      SLRM=AM*PSLM                                  18
```

continued.....

```
C                   COMPARTMENTS
     YF=YF+BRF-DRF-SRF                        19
     YM=YM+BRM-DRM-SRM                        20
     AF=AF+SRF-SLRF                           21
     AM=AM+SRM-SLRM                           22
     TOTSF=TOTSF+SLRF                         23
     TOTSM=TOTSM+SLRM                         24
C                   END OF TIME LOOP
 100 CONTINUE                                 25
```

In this program we use the DATA statement form to supply
the numerical values for initial conditions and constants.
This form involves listing the parameter abbreviations in a
block, followed by the numerical values associated with each
and in the same order as the abbreviations.

The definitions of the stochastic rate processes, the
overall birth rate, and the death rates, are programmed in
the same way as in the CSMP version, but the name for the
function which generates sequences of pseudo-random numbers
is RANDX in this case and the AFGEN function is replaced
by the equivalent TABFN. Notice that the input of numerical
data for each use of the TABFN is not included in the part
of the program given here: the technical details of how this
operation is accomplished are explained in Appendix II. Once
again, a contrast with the CSMP version is that the statements are listed in strict calculation order. For example,
when calculating the overall birth rate the first instruction, in line 4, is concerned with generating the required
random numbers. The definition of overall birth rate, in
line 6, is the last step in the calculation of this rate
process.

Apart from the different function names and the differences associated with effecting the integration procedure
in the DO loop, the FORTRAN version has a close overall re-
semblance to that in CSMP.

Before leaving this example of programming a simple
population model, it should be noted that the way we have
chosen to treat it here, in the same format as that used
for the previous examples of modelling *continuous processes*,
is not the only way in which the essentially *discrete events*
of population dynamics may be handled. On grounds of ease of

model description and economy of computer costs it may be preferable to adopt alternative formats for large models which are concerned exclusively with discrete events and we explore some of those alternatives in Chapter 5. At the same time, as we have seen, small, relatively simple population models fit into the continuous-system mould quite naturally and easily and that format is especially convenient for dealing with 'hybrid' models, which contain both continuous and discrete changes.

EXAMPLE EXERCISES

1. Starting with 60 individuals, a colony of bacteria increases in number by 2 per cent every half-hour. Write a program in CSMP and/or FORTRAN to calculate how many bacteria are present every hour over a twenty-four-hour period.

2. A pasture has an initial weight of herbage of 1200 kg ha^{-1}. If the temperature is 4°C or below, its net growth rate is zero; between 4°C and 25°C the growth rate increases linearly to 150 kg ha^{-1} day^{-1} and then falls off, also linearly, to zero at 35°C and above. Write a program to calculate the changes in weight of herbage present over a ten-day period during which the daily temperatures are as follows:

Day	Temperature (°C)	Day	Temperature (°C)
1	6	6	15
2	2	7	11
3	8	8	28
4	24	9	33
5	29	10	7

3. The pasture defined in Exercise 2 is grazed by cattle and the herbage consumption varies according to the amount present. Consumption is zero if the amount present is 700 kg ha^{-1} or less; between 700 kg ha^{-1} and 1500 kg ha^{-1} the daily consumption rate rises

linearly to 120 kg ha^{-1} and remains constant at 120 kg ha^{-1} for amounts in excess of 1500 kg ha^{-1}. Modify your program for Exercise 2 to include the grazing pro- cess. Calculate the herbage present each day and the total amount consumed by the cattle over the ten-day period.

4. Modify the program for the animal population model given in the text to explore the effects of varying the pro- portion of adult females sold each year by 3 per cent above and below the given value of 29 per cent. What do you conclude about the sensitivity of the system to such changes in this parameter?

COMPUTER PROGRAMMING FOR DYNAMIC SYSTEMS MODELS (II): FURTHER CONCEPTS AND TECHNIQUES

The examples of dynamic models presented in Chapters 2 and 3 were selected to illustrate the model-building process as simply as possible. In this chapter and the next we introduce some further concepts and procedures so as to broaden the range of application.

The following discussion is centred on ideas for modelling particular aspects of biological systems and the techniques for implementing them on the computer rather than attempting exhaustive descriptions of more complex models. Detailed accounts of a range of complete modelling projects are readily available in the literature and it may be more useful to discuss some of the common problems which arise than to present case histories for individual exercises. At the same time it must be appreciated that in a new and rapidly developing field the choice of particular problem areas is necessarily somewhat arbitrary. The reader is urged to give particular attention to the suggestions for further reading at the end of this chapter so as to begin to familiarize himself with the 'working' literature in the field.

4.1. FLOW RATES AND THEIR CONTROLS

In dynamic models of agricultural systems we represent system behaviour in terms of the fluctuating contents of the compartments or 'state variables'. Since those fluctuations are the direct consequence of the physical flows of material into and out of the compartments, it follows that all external and internal controls on the system operate by affecting the flow rates. In other words, the definitions of the flow rates must include all the relevant quantitative information on how the system is affected by its environment and on the ways in which it regulates its own behaviour, internally. Both in terms of analysing how a system works and of devising a computer program to imitate it, the major task is to deal with the flows of material. Having done that job the computation of the contents of the compartments commonly is a relatively simple process.

4.1.1. *Flow rates controlled by one factor*

The simplest possible case is where a rate is controlled by one, constant factor, as in the relative growth model (p.41). There we derived the absolute growth rate (AGR) as the product of the weight of the organism and the relative growth rate (RGR), i.e. the fixed proportion by which the organism is assumed to increase its weight in unit time.

$$AGR = WT * RGR$$

If we make the alternative assumption that RGR is not fixed but varies, say with time, the final calculation of AGR remains the same. But it is no longer appropriate to specify RGR in a PARAMETER instruction and we need to substitute a definition of how RGR varies with time. It is often convenient to insert such a relationship by using an arbitrary function-generator such as AFGEN in CSMP. For example, the following instructions would define RGR as having the value of 0.005 at the start of a 50-day growth period, increasing linearly to 0.01 on the tenth day, to 0.02 on the twentieth day and then remaining constant up to day 50.

$$RGR = AFGEN(TRGR,TIME)$$

FUNCTION TRGR = (0.,0.005),(10.,0.01),(20.,0.02),(50.,0.02)

As we noted in the analogous case of defining the unit rate of photosynthesis in the carbon metabolism model (p.55), the first choice for defining such a relationship is an algebraic equation. The AFGEN function serves to define an approximate relationship where the data is inadequate to derive an equation.[1]

---

[1] The use of simple, linear interpolation between adjacent pairs of points in the AFGEN function is a neutral procedure, with no special assumptions about the form of the curve defining the relationship. For some

continued over.....

4.1.2. Donor controls

Where a flow rate is influenced by the contents of a preceding compartment it is said to be *donor controlled*. If
control is exercised by a succeeding compartment this is
called a feedback loop (see Section 4.1.3). Such controls
may be the only factors involved, or they may be components
of a set of controls on a flow rate. We treat them here as
single factors, as a convenient way of establishing their
effects.

Some influence of the donor compartment on a flow rate
is common in biological systems, with the amount or concentration of material present conditioning the rate of onward
transfer and/or transformation. At the biochemical level,
both catabolic and anabolic processes are generally dependent,
at least in part, on the quantity of the 'raw material(s)'
present. Similarly, the rate of food consumption by animals
is usually controlled, within limits, by the amount of food
available, and so on.

Accounting for donor controls in computer modelling may
be done by specifying either the proportion of material involved in the onward flow per unit time or by focusing on the
'residence time' in a given compartment, commonly called the
'delay time'. An example of the former technique is the handling of the maintenance component of respiration in a model
of the carbon metabolism of plants by Ryle, Brockington, Powell,
and Cross (1973). Total respiration is taken to be made up
of two components: *synthetic* respiration which provides the

....continued

applications, it may be appropriate to use the more elaborate forms of
function generator which are available in CSMP, called NLFGEN and FUNGEN.
The NLFGEN function invokes second-degree interpolation, involving the
three co-ordinates around each point on the curve. When using FUNGEN,
the user may specify interpolation from the first to the fifth degrees.
If it is appropriate to telescope the operations of curve-fitting and
using the fitted relationship, these forms may be convenient. But any
discontinuities in the data can produce unexpected results and it is
good practice to verify that the curve produced is an acceptable approximation by setting up a dummy run of the model to inspect it over the
full range of values. Alternatively, curve-fitting may be carried out
as a separate exercise, using standard procedures.

the energy required for transporting carbon compounds to the sites for growth and the synthesis of more elaborate compounds in new tissue, and *maintenance* respiration which supplies the energy needed to maintain and repair existing tissues. The maintenance component is a charge on the organic material present and is taken as a proportional loss per unit time:

$$MR = WT * MRF ,$$

where MR is the rate of loss, WT the weight of tissue, and MRF the proportion consumed in respiration.

The concept of bottlenecks or *delays* has received considerable attention in modelling industrial and engineering systems (Forrester 1961; Ch.9 and Appendices D and H). We consider here two formulations which are relevant to biological systems; *pipline* and *exponential* delays.

The passage of individuals in a population of organisms through various growth and development stages is one of the biological phenomena which conveniently may be simulated by the use of *pipeline* or *dead-time* delays. This involves an exact reproduction in the outflow from a compartment of the pattern of inflow over time, with a specified lag period or residence time, as illustrated in Fig.4.1.

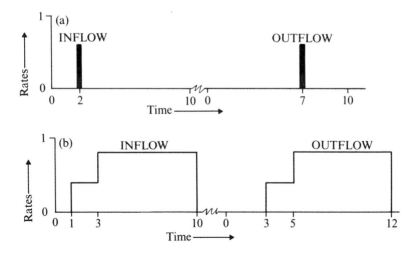

Fig.4.1. Pipeline delays: (a) pulse input, delay period 5 time units; (b) step input, delay period 2 time units.

Programming pipeline delays involves retaining the values for the inflow during the required delay period and then using those values to specify the outflow rate. In FORTRAN, this may be most conveniently accomplished by progressively shifting the values stored in the elements of an array. The special-purpose languages like DYNAMO and CSMP have functions which allow the user to do the same job with less detailed instructions. The 'boxcar-train' feature of DYNAMO allows for the convenient specification of a series of boxes in which to store the required values and to shift them progressively down the series from one box to another. In CSMP, a pipeline delay may be inserted with one instruction only, using the function called DELAY and an example of the use of this function is included in the programme on p.75.

The existence of *decay* processes in biological systems, in which material either literally decays or is dissipated in some other way at a rate proportional to its current amount, makes it relevant to consider another type of delay which has figured prominently in models of industrial systems, the *exponential* delay (Forrester 1961, Chapter 9). Exponential delays involve not only a simple lag period but a change in the pattern of the outflow rate over time compared with the inflow. The basic element of an exponential delay is an outflow rate which is defined by dividing the contents of the compartment by the mean length of the delay period:

$$RTOUT = CPT/DEL \ ,$$

where RTOUT is the outflow, CPT the amount of material in the compartment, and DEL the average delay time.

More than one such element may be included in an exponential delay, with the outflow from the first compartment constituting the inflow to the second, and so on (Fig.4.2). In such 'cascaded' formulations, the divisor to calculate the outflow from each individual compartment is taken as the overall delay period divided by the number of elements in the series. An exponential delay with one element is termed a *first-order* delay, a *second-order* delay has two elements, etc.

The following example program in CSMP illustrates the construction of a first-order exponential delay.

```
        INITIAL                                      1
        PARAMETER DEL=5.0                            2
        DYNAMIC                                      3
            RTIN=IMPULS(0.0,100.0)                   4
            RTOUT=CPT/DEL                            5
            CPT=INTGRL(0.0,(RTIN-RTOUT))             6
            PD=DELAY(50,DEL,RTIN)                    7
        METHOD RECT                                  8
        TIMER DELT=0.1,FINTIM=30.0                   9
```

The CSMP function IMPULS in line 4 defines the rate of inflow to the compartment as a single *pulse* with the value of 1.0, at time zero. This function returns a value of zero

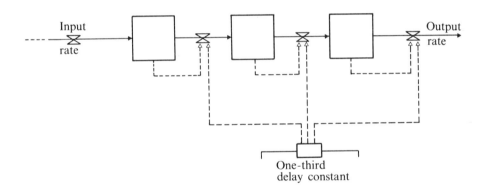

Fig.4.2. Flow diagram for a third-order, exponential delay.

thereafter, the option to generate further pulses being suppressed by setting the second term in the brackets to a value greater than the length of the simulation run, FINTIM.

Outflow from the compartment is calculated, in line 5, as the contents of the compartment divided by the delay period, of value 5.0. Line 6 creates the integral CPT, with an initial value of zero.

Line 7 is included as an example of the use of the DELAY function in CSMP to create a pipeline delay. The first term in parentheses is the maximum number of points sampled during the delay period to characterise the input to the function;

the second and third terms define the length of the delay
period and the expression to be delayed.

The pattern of the outflow rate following a single pulse
input to this first-order exponential delay is illustrated
in Fig.4.3(a).[1] In Fig.4.3(b) the response to a step-change
in the inflow at time zero is shown. The effect of an increase
in the inflow from zero to unity was obtained by deleting line
4 from the program given above and substituting a definition
of RTIN as a PARAMETER value. With the specification of the
initial contents of CPT as zero, the effect of an abrupt increase in the inflow rate at time zero was thus simulated.

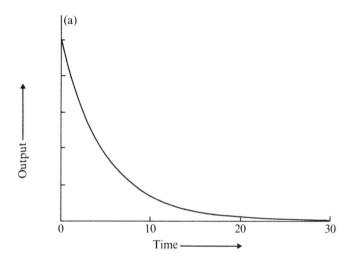

Fig.4.3. Output responses of a first-order, exponential delay: (a) with
a single pulse at time zero.

[1] The graphs in Figs. 4.3 and 4.4 have been produced by using an X-Y
plotting device, such as is available at most modern computing installations. This form of plotting provides a better representation of continuous curves than can be obtained by the use of line-printer plots,
where the representation of small changes is restricted by the fixed
character positions. Use of an X-Y plotter requires alternative instructions to the PRTPLT command in the CSMP program, plus some additional programming tailored to the requirements of the particular installation.

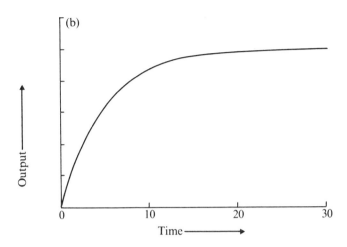

Fig.4.3. Output responses of a first-order, exponential delay: (b) with a step increase at time zero.

By contrast with the output from a pipeline delay the response of the first-order exponential formulation is effectively instantaneous, limited only by the finite time period of the calculation interval in the numerical integration procedure. With the single-pulse input, the maximum rate of output occurs immediately after the input and then gradually declines towards zero. Similarly, the response to a step-change is effectively instantaneous, with an increasingly gradual approach towards equality of the inflow and outflow rates.

Inclusion of more than one element in an exponential delay serves to modify the form of the output response curves so that in the ultimate condition, with an infinite number of elements, the response will be identical to that of a pipeline delay. But for practical purposes the types with a relatively small number of elements are of relevance in modelling biological systems. Figure 4.4 shows the pattern of the response curves with three elements. These graphs were derived by substuting the following instructions for lines 5 and 6 in the program on p.75.

```
DEL3=DEL/3.0
TRANS1=CPT1/DEL3
TRANS2=CPT2/DEL3
RTOUT=CPT3/DEL3
CPT1=INTGRL(0.0,(RTIN-TRANS1))
CPT2=INTGRL(0.0,(TRANS1-TRANS2))
CPT3=INTGRL(0.0,(TRANS2-RTOUT))
```

The responses of the third-order delay do not have the feature of instantaneous response to a change in the inflow rate and the overall form of the curves is close to the expected response pattern of biological decay processes. Where good, detailed data are available, precise curve-fitting procedures may be employed; but the occurrence of exponential-type responses is sufficiently common in biological systems to make the lower-order members of the exponential family useful as first approximations to the expected patterns of behaviour.

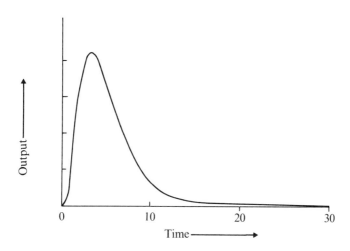

Fig.4.4. Output responses of a third-order, exponential delay: (a) with a single pulse at time zero.

An example of the use of a third-order exponential delay is contained in an analysis of the effects of excreta on the consumption of herbage by grazing cattle (Brockington 1972). Here, the gradual decrease of the inhibitory effect of faeces on grazing in the vicinity of the dung-pats is simulated in this

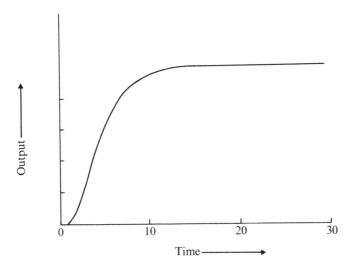

Fig.4.4. Output responses of a third-order, exponential delay: (b) with a step increase at time zero.

way, with the time-constant of the delay estimated from field trials.

Before leaving the subject of donor-control, we may note the use of compartments and their associated outflows as mechanisms to imitate partitioning effects. In the simple model of carbon metabolism in a green plant described in Chapters 2 and 3, the material in the carbohydrate pool is allocated in fixed proportions to the photosynthetic and non-photosynthetic organs. The two outflow rates from the pool are calculated simply by multiplying the contents by the proportionality factors:

$$RT1 = CHP * PRT1$$
$$RT2 = CHP * PRT2$$

The two factors, PRT1 and PRT2, sum to one and since the time step between integrations is unity also, the effect is to remove from the pool in each calculation interval all the material that entered it in the previous interval. Used in this way, the compartment serves only as an arithmetic device for effectively instantaneous allocation and its biological

realism may be open to doubt. At least, the effects of such
a mechanism on overall model behaviour should be checked
before accepting it as a valid approximation. A technical
point which should be noted is that the general form of the
rate equations to redistribute the contents of a compartment
in the next time-step must include the time-step as a divisor.
The first of the two rate equations in the carbon metabolism
model, for example, would be:

$$RT1 = CHP * PRT1/DELT$$

using the CSMP interval variable, DELT, to specify the value
of the calculation interval. It is only in the special case
where DELT = 1. that it may be omitted.

   Substitution of a divisor that is larger than the calculation interval in such an equation will have the effect
of creating a first-order, exponential delay, of course. The
reader may familiarise himself with the effects of this procedure by experimenting with alternative divisors in the
equations for RT1 and RT2 in the carbon metabolism model
and noting the effects on the contents of the pool and on the
growth of the plant parts over time.

4.1.3. Feedback controls
Control of a rate process by the succeeding compartment rather
than the preceding one constitutes a feedback loop. This implies that the contents of the following compartment determine
the flow-rate; the flow rate, in turn, determines the contents of the compartment, and so on, in a circular fashion.

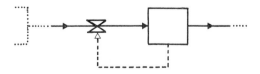

   The relative growth model (p.25) is an example of one
type of feedback loop, the *positive* form, in which an increase
in the contents of the succeeding compartment results in an

increase in the rate feeding into it. Consequently, the
bigger the organism the faster it grows and the whole process is an accelerating one. The curve in Fig.4.5 is derived

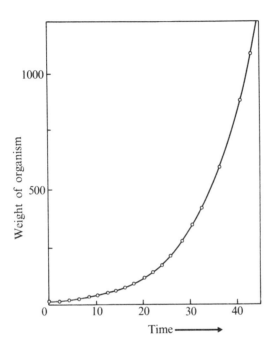

Fig.4.5. Exponential growth with a positive feedback loop: a graph from
the relative growth model with RGR = 0.1.

from the relative growth model on p.42 with the substitution
of the more extreme value of 0.1 for the relative growth rate
parameter, RGR. It clearly illustrates the explosive nature
of an unconstrained positive feedback control.

By contrast with this positive form of feedback control,
a negative-feedback loop involves a regulatory element such
that the rate process adjusts towards an equilibrium. In
man-made or controlled systems it is often called a goal-
seeking mechanism and a familiar physical example is the
governor installed on many machines to regulate the speed
of operation.

Negative feedbacks in biological systems operate in the
same way, making for overall stability by 'damping down'

extreme responses. Examples are the gradual inhibition of
food intake as the contents of an animal's stomach approach
its maximum capacity and the suppression of the production
of new daughter-shoots by the process of tillering in a grass
sward when the number of shoots per unit area approaches
a limiting value.

The contamination of pasture by the excreta of grazing
cattle and the consequent inhibition of herbage intake referred
to above (p. 78) provides another instance of control by a
negative-feedback loop. In that system the amount of faeces
produced is related to the herbage eaten and one of the controls on the rate of intake is the amount of faeces already
produced. Thus a high level of intake will lead to a high
level of contamination of the pasture and a reduction in the
rate of intake. The effect is not a permanent one, because
the inhibitory effect of faeces is gradually dissipated and
in time the intake rate will tend to increase again, only to
be reduced once more as the level of faeces builds up again.
Similarly, an animal's appetite will tend to recover as its
stomach contents are digested and the indigestible material
passes along the gut to be voided as excreta. It is typical
of the many negative feedback controls in biological systems
that their effects are thus temporary in nature, acting as
regulators to prevent extreme behaviour.

Negative-feedback loops are widespread in biological
systems and contribute largely to their inherent general
stability. However, their effectiveness can vary, just as it
does in man-made systems. An inefficient governor mechanism
on a machine can respond too late and/or too violently to the
changes in speed which it is designed to control. The result
is a series of sharp and repeated fluctuations above and
below the target speed, repeatedly over-reacting first in
one direction and then in the other. Such inefficient regulatory mechanisms are relatively rare in biological systems;
but they can occur with some predator/prey interactions for
instance, leading to major fluctuations in the populations
concerned (de Wit and Goudriaan 1974).

The requirements for an efficient control mechanism are
that it should respond to deviations from the target or

equilibrium condition relatively rapidly and that the degree of response should be tailored to the amount of the deviation. This implies that successively smaller corrections are required as the target is approached, a condition which corresponds to an exponential-type curve. Mathematically, we find that a simple, first-order negative-feedback loop produces such a curve. This is because in a closed loop the output 'becomes' the input. This implies a shape of curve which is not altered by integration, a condition which is satisfied by one of the exponential family of curves. The exponential function remains unchanged when integrated except for a constant multiplier term.

The negative exponential function, $\exp(-t/T)$, when integrated becomes $-T \exp(-t/T)$; where exp is the base of Naperian logarithms, t is time and T is the time constant of exponential change.

Similarly, the positive exponential function $\exp(t/T)$ yields, on integration, $T \exp(t/T)$. Thus the first-order positive-feedback loop generates an exponential response curve also (Fig.4.5). But in this case there is an increasingly divergent response, of course, in contrast to the convergence towards an equilibrium value with the negative form.

At first sight, the utility of the positive-feedback loop in modelling biological systems appears very limited. The number of situations in which growth or accumulation of material exactly follows an exponential form is indeed very small. But there are many instances where it is both relevant and convenient to recognize the basic tendency to the exponential form and then to build on the relevant constraints as modifying factors. Reproduction in a population of organisms frequently has this basic form because of the link between the number of adults existing at a given time and the rate of reproduction. Representing the *potential* rate of increase in numbers in this way, the *actual* rate is then calculated by allowing for such constraints as the natural mortality rates and the offtake or selling rates in populations managed for agricultural purposes.

Both positive- and negative-feedback loops are usually simple to incorporate in a computer program by applying appropriate coefficients; but a technical point which may require some attention is the specification of the time constant of the exponential curve in relation to the calculation interval in the numerical integration procedure. In Section 3.4 we explored the effects of different calculation intervals on the solution of the relative growth model, compared with an exact, analytical solution. That comparison assumed the true, 'instantaneous' value for the coefficient, RGR, was known, a condition which is seldom true in practice of course. More commonly, we have an estimate of the coefficient over a finite period of time and the choice of an appropriate interval between successive integrations should be related to that time period between the successive experimental observations. With a positive feedback loop, an integration interval equal to the period between observations will generally suffice to represent the behaviour of the system *within the limits of the available data*. Where a negative feedback is to be modelled, the integration interval should never exceed the measurement interval. A larger interval may introduce not only computational errors, but it can result in an artificial, oscillatory pattern, unrelated to the inherent behaviour of the system. (See Forrester 1968, Section 6.3, for a detailed discussion of the choice of integration interval in relation to negative feedbacks.)

4.1.4. Control of flow rates by more than one factor
Analysis of biological and agricultural systems frequently reveals rate processes which are controlled by a number of factors, either state variables within the system or analogous driving variables outside its boundary. Commonly, the effects of the individual controlling variables are not *independent* in the statistical sense and this means that in experiments to establish their effects it is necessary to explore their combined effects in all the various possible permutations. If such comprehensive, factorial-type experiments have been carried out, it is relatively simple to incorporate the results in a model; but the technical difficulties of carrying out the required experiments often result, in practice, in incomplete and fragmentary information.

In principle, multi-factorial control of a rate process
is an analogous situation to that discussed in section 1.4.1
for models of whole systems. The modeller may treat the flow
rate and its set of controls as a single, black-box element
within the model, or an attempt can be made to split it up
and analyse it in greater detail on a more mechanistic basis.
It is often possible to break down the overall rate into a
series of successive steps, with the objective of defining
exactly where and how each of the control variables exerts
its particular influence. Figure 4.6 illustrates the principle of such a subdivision and shows how it may be possible,

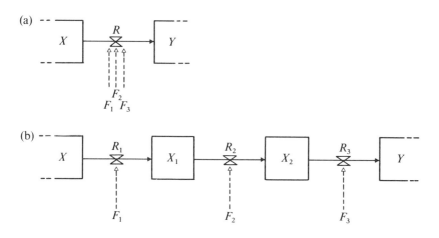

Fig.4.6. Alternative model formulations for a rate process influenced by
three controlling variables. In (a) the overall rate is taken as a
single element and in (b) it is resolved into three stages, each controlled by a single variable.

in theory, to facilitate the characterization of the effects
of the control variables by breaking them down into more
manageable subsets, if not into single factors. But consideration of real-life examples of subdividing overall rate
processes, such as photosynthesis in plants and digestion of
food in the alimentary tract, emphasizes that the potential
advantages may be offset by the much greater volume of information required to support a detailed breakdown and
analysis. If the treatment of an overall rate process as a

single element is in line with the general level of resolution required to satisfy the objectives of a particular
modelling exercise it may not be appropriate to analyse it
in much greater detail.

Nevertheless, the principle of considering the component
processes within an overall flow rate may be a useful conceptual aid in analysing multi-factorial controls. As Rabinowitch (1951) has pointed out in a discussion of the kinetics
of the rate of photosynthesis, the widely used framework for
analysing such controls which was proposed in slightly different terms by Liebig (1840) as the 'law of the minimum' and
by Blackman (1905) as the 'law of limiting factors', must
imply a sequence of processes within the overall flow rate.
For example, it is only in the circumstance that Blackman's
'slowest' or 'limiting' factor acts on one of a series of
stages within the overall process, thus creating a 'bottleneck', that the overall rate may be expected to be independent
of factors which influence other processes in the sequence.
If the flow rate is a single reaction process, its rate normally will be a function of all of the controlling variables.
Thus in the 3-stage process of Fig.4.6(b) it is possible to
derive response curves which approximate to the classic type
postulated by Blackman as in Fig.4.7(a). Over the section
A-B, the overall rate is controlled only by factor F_1, acting
on the first stage transfer, R_1. The plateau B-C is imposed

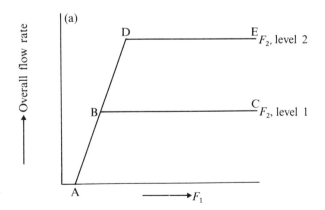

Fig.4.7(a). Blackman-type response curves, with factors F_1 and F_2 influencing separate stages in the overall flow rate.

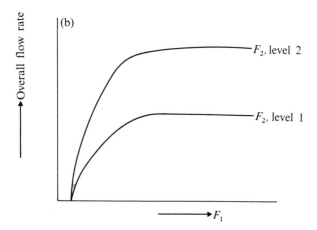

Fig.4.7(b). Response curves with factors F_1 and F_2 influencing the same stage in the overall flow rate and maxima controlled by F_2.

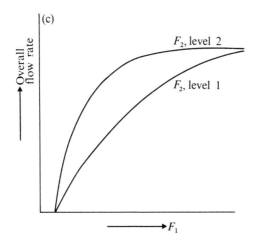

Fig.4.7(c). Response curves with factors F_1 and F_2 influencing the same stage in the overall flow rate and maximum response conditioned by a third factor, F_3.

by the factor F_2 setting a limit on the overall rate at the second stage, R_2. If the limit on R_2 is raised by increasing

F_2 to a higher value, this allows the further dependence of the overall rate on F_1 to be manifested over section B-D, followed by the plateau D-E as R_2 once more constitutes the limiting stage.

Major features of classic, Blackman-type response curves are the coincidence of the initial slopes and the abrupt transitions to the various plateaux. The derivation of curves which correspond exactly to this pattern depends not only on the existence of a series of stages within the overall rate process and their separate control by different factors, but also, for instance, on the different stages having widely different maximum rates. Rabinowitch (1951; Ch. 26 and 27) provides a comprehensive review of the kinetics of multi-stage processes, including the detailed effects of various combinations of controls. The exact forms of overall response are, of course, of critical importance if they are to be used as an analytical tool to deduce the mechanisms within an overall rate process. Here we are concerned with some of the gross differences which result from different arrangements of the controls as an aid to formulating an acceptable approximation to the behaviour of the overall rate process.

If the variables F_1 and F_2 affect the same stage, say R_1, then the initial slopes of the response curves will diverge rather than coincide. Divergence of the initial slopes may be combined with a series of separate plateaux, due to successive limits set by F_2 acting on R_2, as in Fig.4.7(b). Alternatively, a single plateau may be imposed by the influence of another variable acting on a different stage, e.g. F_3 acting on $R3$ (Fig.4.7(c)).

In programming multi-factorial controls involving limiting factors or processes it is often convenient to use a function for selecting a *minimum* value from a number of variables. For example, Fig.4.8 illustrates the response of an overall flow-rate, R, to variation in two factors F_1 and F_2. For purposes of illustration, the potential response to both factors over their maximum ranges is assumed to be linear and equal. The actual response of either of the two factors may be limited by the other, depending on their relative values. For instance, if F_1 has a value midway

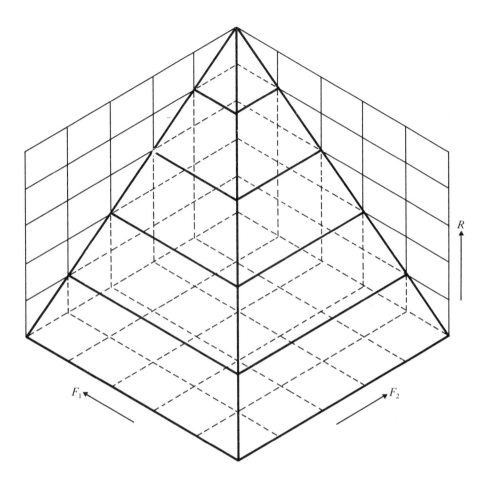

Fig.4.8. Response surface of an overall flow rate R to two variables, F_1 and F_2, on a limiting-factor basis.

between the minimum and maximum points of its range, the overall flow rate will be dependent on F_2 when F_2 varies between
its minimum and mid-points. Increasing the value of F_2 beyond
the mid-point will have no effect on R while F_1 remains at
half its maximum value. Conversely, fixing the value of F_2 at
a given level will impose a 'ceiling' on the response to F_1.
Having ensured that the absolute numerical values of F_1 and
F_2 are expressed on a common scale, the key operation in calculating their combined effect on R is to identify which is
the smallest on that scale, i.e. to select the minimum.

The following CSMP code uses the device of selecting a minimum numerical value to calculate the symmetrical response surface of Fig.4.8.

```
XF1=AFGEN(TF1,F1)                               1
FUNCTION TF1=(0.0,0.0),(10.0,1.0)    2
XF2=AFGEN(TF2,F2)                               3
FUNCTION TF2=(0.0,0.0),(20.0,1.0)    4
REDF=AMIN1(XF1,XF2)                         5
R=POTR*REDF                                        6
```

In this example, the variables F_1 and F_2 are taken to have numerical values ranging between 0-10 and 0-20 respectively. The AFGEN functions in lines 1-4 convert those numerical ranges to a common scale of 0-1 for the derived variables XF1 and XF2. The minimum of XF1 and XF2 is selected in line 5[1] and this is used in line 6 as a multiplication factor to compute the actual flow rate from its maximum, potential value, POTR.

Variations in the form of the response surface from the above example, with its equal, linear effects, may be effected by appropriate modifications to the AFGEN functions or to their equivalent, algebraic equations. It is possible, also, to incorporate the selection of a minimum value from more than the two variables used in the example. The device of defining an overall maximum value for a flow-rate and subsequently constraining it by the application of one or more 'reduction factors' is a fairly common one in physiological modelling (e.g. de Wit and Goudriaan 1974).

A more general format for defining the response of a flow rate to two controlling variables in CSMP is the TWOVAR function. This is analogous to the AFGEN function and builds up an approximate picture of the response surface by the specification of a number of 'slices' in one dimension and allowing for linear interpolation both within and between the slices. This format, or an equivalent algebraic definition of the surface, does not necessarily include any element of

---

[1] It should be noted that the exact form of the library function for selecting a minimum value in CSMP and FORTRAN varies according to whether the inputs and outputs are real or integer variables.

limiting factors, of course.

4.2. TRANSPORT PROCESSES

An important feature of many agricultural systems is the movement of materials from place to place: gases like oxygen and carbon dioxide, water in liquid and gaseous form, and dissolved solutes. The movements of water in soils and crops in response to potential energy gradients, the transport of ions in the water by mass flow, and their movement by diffusion, and the diffusion of gases to and from the sites of biochemical processes are some of the many transport processes which have a major influence on the behaviour of biological systems.

These movements are governed by fundamental physical laws and for simple situations it may be possible to solve the appropriate equations analytically. But simple situations are the exception rather than the rule in agricultural systems. For instance, the movement of ions in the soil solution is frequently complicated by adsorption and desorption and by uptake by plant roots and micro-organisms (Frissel and Reiniger 1974). In such cases the use of what may be called 'layer models' generally has proved useful (de Wit and Van Keulen 1974; Radford and Greenwood 1970; Hillel 1977). The total transport path is divided up into a number of layers or slices, each represented by one of a chain of compartments. The overall flow is then approximated by considering it as made up of the series of finite steps involved in the transfers between successive compartments.

This is a convenient method of simulating transport processes in biological systems because it allows the formulation and solution of more complex problems than could be attempted otherwise. But it should be appreciated that this is one of the relatively few areas in bioloogical modelling where the approximations of numerical methods of integration can present technical problems. The reader should consult the references cited above for a detailed treatment of these technical aspects. Here, we consider a simple example of soil-water flow to illustrate the principles of the method.

When water from rainfall or irrigation arrives at the

soil surface a proportion may be lost by surface run-off and the balance will enter the soil and move downwards. As a first approximation this major downward flow may be simulated by considering a number of layers within the profile, each with a maximum water-holding capacity (the 'field capacity' condition), so that surplus water will 'overflow' from one layer to the next one below until it reaches a layer which is incompletely filled (Brockington 1971). For agricultural purposes, the part of the profile containing the crop roots is of major interest and the overflow from the lowest layer within the rooting zone is taken as the rate of soil 'drainage'. In addition to this drainage loss, water is removed from the soil system by the crop roots to make good the loss from the aerial parts by transpiration and there may be a component, also, of direct evaporation from the surface layer of the soil.

The dynamics of soil water movement thus may be accommodated by a series of compartments as shown in Fig.4.9, each of which represents the water in one layer within the total profile. The uppermost layer receives water from rainfall and/or irrigation and the lowest one is the source of any internal drainage. Otherwise, each layer will receive water from that above it, will pass on any surplus to its immediate neighbour below and may lose water by uptake by the plant roots. Accordingly, the balance of water in a typical layer may be represented by the integral equation:

WCONT = INTGRL(IWC,(INF-OUTF-UPT)),

where the water content depends on its initial level (IWC), plus the addition of water from the layer above (INF), and minus the transfer to the layer below (OUTFL) and the uptake by the plant roots (UPT).

The above treatment of soil water dynamics is very much an approximate one, depending on a number of simplifying assumptions, e.g. the absence of downward flow through macroscopic channels, which may be more or less independent of the water content in the soil crumbs. In addition, the depth of the layers considered and the solution interval need to be appropriately balanced. The majority of dynamic soil-

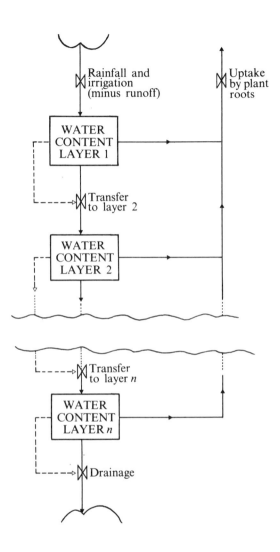

Fig.4.9. Flow diagram of a simple soil-water model.

water models have been constructed on a more detailed basis, incorporating the basic transport equations in place of more empirical parameters such as field capacity (e.g. van Keulen 1975). But the basic feature of dividing up an overall flow path into a number of segments is a common one in modelling transport processes at various levels of resolution.

In programming layer models, an important common charac-

teristic is the repetition of the same set of calculations
for all the layers, with the general exception of the first
and last in the sequence. The need to carry out the same
calculations for a number of units which differ only in the
numerical values of the inputs arises in a number of other
areas of biological modelling, e.g. in population models
concerned with a number of individual organisms or similar
groups of individuals. It can be both tedious and conducive
to programming errors to repeat the same set of instructions
for each unit and one way of dealing with the situation is
to make use of the *Macro* feature which is available in lan-
gauges like DYNAMO and CSMP. This allows the user to specify
the calculations once only and then to call on that set of
instructions a number of times, as required.

According to the nature of the numerical inputs to the
repeated calculations the same purpose may be served in
FORTRAN either by executing the repetitions within a DO loop
or by defining the calculation structure in a Function or
Subroutine. If the numerical outputs from one element form
the inputs to the next in the series, as they do in many
layer models, the DO loop is a convenient choice - as it is
for the analogous task of the repetitive calculations involved
in numerical integration. If independently derived numerical
inputs to each element are required, the Function or Subroutine structure generally will be appropriate.

The use of a Macro element in CSMP is essentially equivalent to a Function or Subroutine, with the added flexibility and convenience of the availability of the CSMP special
functions for integration and other tasks in making up the
package of instructions. The package is inserted at the
beginning of the program and is demarcated by a first line
containing the word MACRO and a final instruction, ENDMACRO.
In between these two lines, the calculations to be carried
out on each similar element in the model are specified. The
MACRO may be called on, or 'invoked', by a single instruction
each time it is required in the remainder of the program.

The following example of using a MACRO in CSMP is based
on its application in a model of the carbon metabolism of a
grass plant to define the calculations applicable to each of

the several leaves on the plant (Ryle *et al*. 1973). Each of
the leaves fixes carbon dioxide and a proportion of the carbohydrate produced is expended in synthetic respiration to meet
the energy requirements of translocation to the growing regions
and the elaboration of the compounds required for the new
tissues.   The balance of the carbohydrate is allocated to the
growing points.

```
*     MACRO DEFINING CALCULATIONS FOR ALL LEAVES
MACRO A,B,C,D,E,F=LEAF(DN,P,LA,R,TA,TL,TS,TR)                    1
      RPH=P*LA*0.682*DN                                         2
      A=INTGRL(0.,RPH)                                          3
      CHP1=INTGRL(0.,(RPH-RRESP-DRPH))                          4
      RRESP=CHP1*R/DELT                                         5
      B=INTGRL(0.,RRESP)                                        6
      DRPH=CHP1*(1.,-R)/DELT                                    7
      CHP2=INTGRL(0.,(DRPH-TRAM-TRL-TRST-TRRT))                 8
      C=CHP2*TA/DELT                                            9
      D=CHP2*TL/DELT                                           10
      E=CHP2*TS/DELT                                           11
      F=CHP2*TR/DELT                                           12
ENDMACRO                                                       13
INITIAL                                                         .
  .                                                             .
  .                                                             .
DYNAMIC                                                         .
                                                                .
*     CALLING MACRO FOR LEAF ONE AND SUPPLYING INPUTS
      A1,B1,C1,D1,E1,F1=LEAF(DN1,P1,LA1,R1,TA1,TL1,TS1,TR1)50
      P1=AFGEN(TP1,TIME)                                       51
FUNCTION TP1=(0.,350.),(2.,470.), etc.                         52
```

Line 1 of the above code indicates that a MACRO element
is to be defined, with the specific name LEAF, with the numerical inputs to the calculations as listed within the parentheses
and with the required outputs as specified in the list to the
left of the equals sign.
      Lines 2-12 specify the calculations required each time
the MACRO is invoked and line 13 signifies the end of its
definition.
      Line 2 defines the rate of photosynthesis as the pro-
duct of the rate per unit area (P), the total leaf area (LA),
the factor 0.682 to convert units of carbon dioxide to carbohydrate and a trigger (DN).   The variable DN is defined elsewhere to act as a 'clock', having a value of one during the

daylight period and zero during the dark period in each 24-hour cycle. Total photosynthesis is summed, for reference purposes, in line 3. The integral of line 4 (CHP1) is the first of two carbohydrate pools. This first pool receives the product of photosynthesis and allows for the deduction of the synthetic component of respiration. The rate of synthetic respiration is defined in line 5 and summed in line 6. The second pool of carbohydrate (CHP2) receives the balance of carbohydrate from CHP1 and allows for its translocation to the apical meristem (TRAM), to the other leaves on the plant (TRL), to the stem (TRST) and to the roots (TRRT). The rates of translocation are defined in lines 9-12. (In the full, working version of the program for this model some further calculations are included in the macro definition, not shown here).

An example of calling on the MACRO is shown in line 50. Notice that both the input and output variables are specifically identified as those relevant to this particular invocation of the MACRO for leaf one. Lines 51 and 52 give an example of defining the numerical values of the inputs for this particular call on the MACRO. The output variables, A1-F1, are available from the results of the calculations within the MACRO for use elsewhere in the program.

The reader should refer to the CSMP manual for detailed instructions on the use of the MACRO facility and for a discussion of its use for various applications compared with alternatives such as Subroutines and Functions.

In concluding this chapter, it is reiterated that the reader should aim to supplement the limited range of topics discussed by exploring the working literature on simulation modelling. The references cited under the heading of suggestions for further reading (p.148) are designed to supplement those cited earlier to facilitate making a start on this task.

A further vital element for the serious student of this form of modelling is *practice*! There is no substitute for practical work to provide a realistic insight into what it can and cannot do. Some suggestions for practical exercises are given below and the student is urged to devise variations and additional topics for himself in his own fields of special interest and experience.

SUGGESTED EXERCISES

1. Experiment with divisors greater than 1 in the equations for RT1 and RT2 in the carbon metabolism model on p.54. Add an assumed negative-feedback effect on the rate of photosynthesis when the level of material in the carbo-hydrate pool exceeds given values.

2. Construct programs to explore the output responses of exponential delays with the number of elements varying between one `and six. Use pulse inputs and stepwise in-creases and decreases in the input rate.

3. Write a program to simulate a transport process in a biological system, using a Macro element or related device, as appropriate.

4. Simulate the growth (or some other characteristic of interest) of a number of individuals in a population, again using a Macro element or similar formulation.

COMPUTER PROGRAMMING FOR DYNAMIC SYSTEMS MODELS
(III): EVENT-ORIENTED MODELS

5.1. STRUCTURE OF EVENT-ORIENTED MODELS

In previous chapters we have been concerned with *continuous* simulation models, in which time is incremented in small steps to approximate the continuous changes which are characteristic of many biological systems. As we have seen, this type of model can also accommodate *discrete* changes or *events* and it may be the most convenient way of representing systems in which there are both continuous and discrete changes. In this section we introduce the major alternative formulation for dynamic models, in which time is advanced in *variable increments*, from one 'instantaneous' event to the next in the time sequence. Event-oriented models are of special relevance in modelling, for example, the behaviour of populations of organisms. The birth and death processes in populations may be regarded as instantaneous events and the progression through a series of development stages also may be treated as a series of finite steps.

We need to distinguish the use of variable time-increments in event-oriented models from the variable *calculation intervals* which are used in some numerical integration procedures to mimic continuous changes (Section 3.4). The device of incorporating a variable period between successive iterations of a numerical integration procedure is designed to effect a compromise between accuracy and computer usage in the simulation of continuously changing variables. In event-oriented models we are not concerned with continuous changes and the varying time increments arise from skipping over the periods between successive, instantaneous events.

Where the sequence of discrete changes is an irregular one, it might be expected that moving directly from one event to the next in the time sequence might be more economical of computing time than the procedure of advancing through time in the small, regular steps in continuous formulations. This potential economy may not be realized in practice, however, as overall computer use depends partly on other features in the organization and running of the programs (Conway, Johnson, and Maxwell 1959). A more important contrast with continuous models is that discrete formulations are usually concerned more with analysing system behaviour on a 'statistical' basis than with following the exact sequence of individual changes in chronological order.

Historically, event-oriented computer modelling has been developed chiefly to analyse such systems as the movement of customers through retail sales outlets, the servicing of ships and their cargoes in port installations, and the handling of information and instructions in data-processing operations. Interest in this type of system tends to focus on aspects such as the time spent waiting in queues and the periods occupied by the 'servicing' or 'processing' of items passing through the system (Baker and Dzielinski 1960; Blake and Gordon 1964). Although the individual computations in a discrete simulation run must be carried out in the correct overall time sequence, the results are normally presented as summary statistics for residence periods of items in the various parts of the system, irrespective of the chronological order of individual events (Efron and Gordon 1964). This form of output contrasts markedly with the serial record of individual changes which forms the primary output of continuous models - although it is possible to abstract similar summary statistics from the latter as an additional operation.

5.2. COMPUTER LANGUAGES

A number of languages have been devised for programming event-oriented models. These include the IBM General Purpose Systems Simulator, 'GPSS' (IBM 1971); the closely similar Xerox General Purpose Discrete Simulator, 'GPDS' (Xerox Corporation 1972); and SIMSCRIPT (Hanser, Markowitz, and Karr 1962). Reviews of these and other languages are given by Krasnow and Merikallio (1964) and Tocher (1965). GPSS must be rated amongst the most versatile of these languages and is used in the following example as an illustration of the construction and use of discrete models.

5.3. AN ANIMAL POPULATION MODEL PROGRAMMED IN GPSS

Our example is based, with some minor modifications, on the
animal population model which was given in a continuous
formulation in Section 3.6. In this version, the birth rate
is taken as 60 per cent of the adult females in the population. The interval between births ranges from 335 to 395
days. The sex ratio at birth is 48 males to 52 females and
the probabilities of death for males and females in the first
year of life are 7 and 5 per cent, respectively. All the surviving males and 40 per cent of the female stock are sold at
one year of age and the remaining females enter the breeding
herd. The females are culled after seven breeding cycles.
In this form, the model would correspond generally to the
operation of an extensive, 'ranching' enterprise with beef
cattle, in which there is a relatively low rate of reproduction and the saleable outputs are yearlings and cull cows.

Readers wishing to become familiar with the details of
programming in GPSS in order to construct models for themselves will need to consult the users' manual. In this instance, two versions of the manual are available, an introductory one (No. SH20-0766) and a complete description in the
reference version (No. SH20-0851-1). The introductory manual
is designed for those without previous experience of this
form of programming and contains sufficient basic information for the construction of simple models. The full manual
is essential for reference purposes and for the construction
of more sophisticated programs exploiting the wide range of
facilities in this language.

The program listing of the animal population model in
GPSS given below contains two sets of reference numbers.
Those on the right-hand side serially reference all the statements in the program, while those on the left-hand side refer
only to the *blocks* – which constitute the 'core' operations,
controlling the movements of items through the system. Fig.
5.1 is a diagrammatic representation of the alternative
routes which may be taken by items under the control of the
blocks.

In describing the program we consider first the overall
structure, as illustrated in Fig. 5.1, and then go on to a

BLOCK NUMBER	*LOC	OPERATION	A,B,C,D,E,F,G,H,I	STATEMENT NUMBER
		DRM	FUNCTION	RN1,D2	1
		.07,0/1.0,1		2
		DRF	FUNCTION	RN2,D2	3
		.05,0/1.0,1		4
		YBH	TABLE	P1,1,1,10	5
		SIMULATE		6
1		GENERATE	1,,,100,,2,H	7
2	REC	TRANSFER	.480,FEM,MALE	8
3	MALE	ADVANCE	365	9
4		ASSIGN	2,FN&DRM	10
5		TEST E	P2,1,TDM	11
6		TRANSFER	,TSM	12
7	FEM	ADVANCE	365	13
8		ASSIGN	2,FN&DRF	14
9		TEST E	P2,1,TDF	15
10		TRANSFER	.400,RFM,TSF	16
11	RFM	ASSIGN	1,0	17
12	BRF	TEST LE	P1,6,TCF	18
13		ADVANCE	365,30	19
14		INDEX	1,1	20
15		TABULATE	YBH	21
16		TRANSFER	.400,BIRTH,BRF	22
17	BIRTH	SPLIT	1,REC	23
18		TRANSFER	,BRF	24
19	TDM	TERMINATE	1	25
20	TSM	TERMINATE	1	26
21	TDF	TERMINATE	1	27
22	TSF	TERMINATE	1	28
23	TCF	TERMINATE	1	29
		START	1500,,100	30
		END		31

more detailed exposition of the form and effects of individual statements.

5.3.1. *Routes of transactions through the system*
The items moving through a system are known as *transactions*
and in this example the transactions are equated with animals
in the population. Essentially, animals are born and having
been divided into males and females they move through the
system into different age classes. At appropriate points
in the sequences representing the ageing process in both males
and females, individuals are diverted from the main 'flow'
to account for deaths and sales. In the female sequence there
is a further diversion to account for the culled females.
When the system is 'in motion', all the births originate
from the SPLIT block, numbered 17 in the block series, which
serves to create daughter transactions from the mother transactions passing through it. But, initially, the GENERATE
block (1) is required to simulate a number of births to set
the system going. Whether originating from the SPLIT or
GENERATE blocks, the newly born animals are divided by sex

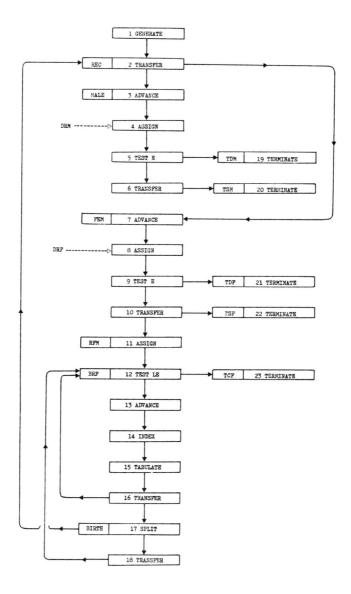

Fig.5.1. Routes of transactions in the animal population model.

in the following TRANSFER block (2) and males are routed to
the sequence in blocks 3-6 and females to the sequence numbered 7-18. In the male sequence, the processes of ageing,
dying, and eventual sale are simulated. In the female sequence, these processes are simulated also, plus the birth
of young and the culling of females after their seventh

reproductive cycle. The sequence of TERMINATE blocks numbered 19-23 serves as the ultimate destination for animals which die, are sold, and culled. At a TERMINATE block the transactions are 'destroyed' by the program to represent the conclusion of their route through the system.

5.3.2. Function of individual program statements
At the start of a simulation run no transactions are present in the system represented by the model and a GENERATE block is required to create transactions, at least to start the overall operation of the model. In this example, with a self-maintaining herd of animals, further transactions are eventually created by the SPLIT block (17). In other applications it may be convenient to create all new transactions by one or more GENERATE blocks. Here, this creation is confined to a hundred animals, with one individual arriving each day in the first hundred days. This 'initialization' of the model might be rationalized as the purchase of 'foundation stock' for a new enterprise; but while this is plausible it obscures the more important point that with the principal analysis of the results focused on how the system operates as a 'going concern' the technical details of how it is set in motion will normally be of only subsidiary interest.

The format for the specification of the GENERATE block conforms to a structure common to all block types in GPSS. Three *fields* are involved in a block definition: the *location* field in columns 2-6, the *operation* field in columns 8-18, and the *variable* field commencing in column 19. The location field is optional and used to define a 'label' by means of which transactions may be routed to the block concerned from elsewhere in the program. The operation field specifies the general type of block and the variable field defines the details of how it is to operate. The variable field is divided into a maximum of nine *subfields*, separated by commas and known as subfields A to I. In any or all of the subfields an *operand* appropriate to the type of block may be inserted to define a particular aspect of the operation of the block. In the definition of the GENERATE block in the example, subfield A is used to specify an interval of one time-unit

(taken as one day for our purposes) between successive transactions created by the block. Following three commas to de-
limit the blank or 'default' specifications in subfields B
and C, the number 100 in subfield D indicates that a total
of one hundred transactions are to be created. Finally,
in subfield F, the operand 2 defines the number of *parameters*
which are to be associated with each transaction created.

The association of parameters with individual transactions is an important device which allows for properties
of a transaction to be defined and used as a basis for
decisions as it moves through the system. In our example
program the first parameter, P1, is used to keep a record
of the number of breeding cycles of each animal in the
breeding herd, so that individuals may be culled at the
appropriate time. Parameter P2 is used in programming the
deaths of animals in their first year of life.

The TRANSFER instruction of block 2 divides the newly
born animals into males and females. As these originate
from the SPLIT block (17) in addition to the preceding
GENERATE block, the label REC is placed in the location
field so that the daughter transactions from the splitting
operation may be routed to this block. In the form given
here, the TRANSFER block effects what is known as a 'statistical' transfer. On average, 480 of every 1000 transactions
entering it (subfield A) are directed to the block labelled
MALE (subfield C) and the remainder to the block labelled
FEM (subfield B).

The first of the blocks in the sequence for male animals,
the ADVANCE instruction of block 3, introduces a time 'delay'
of 365 days - simulating the retention of these animals for
one year after birth. The ASSIGN block (4) is used to set the
value of parameter P2 associated with the males to the output
of the FUNCTION DRM. In the form specified in statements 1
and 2 at the beginning of the program, this FUNCTION returns
a value of zero for insertion in P2 for the 7 per cent of
males dying in their first year and a value of one for those
surviving. The FUNCTION statement in GPSS is basically equivalent to the AFGEN facility in CSMP; but it is possible to
use it either in 'continuous' form, with linear interpolation

between successive co-ordinates, or in a 'discrete' or 'stepped' form. The latter option is chosen in this instance, so that the output is zero when the random-number generator, RN1, takes any value between 0.0 and 0.07 and there is a step-change to an output of one for values of RN1 between 0.07 and 1.0.

Having set the value of parameter P2 in the ASSIGN block, it is used in the TEST block (5) to separate the young males dying from those surviving. In the form here, the test is for equality of the arguments listed in subfields A and B of the variable field. Thus if P2 = 1 transactions pass to the next sequential block, the TRANSFER instruction of block 6. Where the equality is not true, the transaction concerned is directed to the block labelled TDM, as specified in subfield C. The label TDM is attached to the TERMINATE statement in block 19, which forms one of the series of these commands to 'destroy' transactions in the program. As with other block types, a record is kept of all transactions entering each TERMINATE block and this provides a convenient summary of total deaths in the results of a model run.

The TRANSFER statement of block 6 is used in the 'un-conditional mode', with subfield A blank and with the label of the block to which *all* transactions are to be directed given in subfield B. All the surviving males thus are diverted to TERMINATE block 20 which simulates the sales of yearling males. This completes the sequence for the male animals.

Blocks 7, 8, and 9 of the sequence for females are equivalent to those of 3, 4, and 5 in the male sequence: accounting for the passage of time in the first year of life and the deaths during that period.

In block 10, the statistical TRANSFER block is used to divert the 40 per cent of young females which are sold to the TERMINATE block 22.

The ASSIGN statement of block 11 serves the technical purpose of ensuring that the value of parameter P1 for young females is set to zero when they enter the breeding herd. This parameter is used later in the program as a counter to indicate the number of breeding cycles of each individual in the herd so that they may be culled at the appropriate stage. The precaution of setting P1 to zero is required only for those animals originating from the SPLIT block, whose parameter values are taken as identical with those of their parent transactions; all parameter values for transactions created by GENERATE blocks are taken as zero initially.

The TEST block (12) effects the culling of animals after they have completed seven breeding cycles. Each time an animal passes through blocks 12-18, the INDEX block (14) increments the counter in P1 by one. When the counter attains a value of 7 it will fail the test of P1 ≤ 6 and the animal concerned will be directed to TERMINATE block 23, using the label TCF.

The ADVANCE block (13) is coded to account for the variable period of the breeding cycle, with an equal probability of any (integer) number in the range 365 ± 30 days.

The TABULATE block (15) is used in conjunction with the TABLE specification in statement (5) to gather data on the number of breeding cycles of animals in the herd and print out a frequency table in the results from the simulation.

The 40 per cent of animals which fail to produce offspring in each breeding cycle are returned to block 12 from the statistical TRANSFER statement in block 16 and the remainder pass to the following SPLIT block. The splitting operation is coded so that one daughter transaction is derived from each mother transaction entering the block (subfield A) and the daughter transactions are directed to block 2, using the label REC. Parent transactions pass from the SPLIT block to block 18, where they are directed to reenter the breeding sequence at block 12 by the unconditional TRANSFER command using the label BRF.

Block 18 is the last of the sequence simulating the behaviour of females in the herd, with all the animals entering it at block 7 accounted for in terms of deaths, sales, embarking on a further breeding cycle, or being culled. The individuals born during the female sequence are accounted for by being directed to block 2 for partitioning according to sex.

The TERMINATE statements of blocks 19 to 23 have been

referred to above as the destinations of the various classes
of animals leaving the system and it remains to describe the
function of the overall program controls in statements 6 and
30.

The SIMULATE instruction of statement 6 indicates that
the program is to be both assembled and executed: the option
to assemble only, by omitting this command, is designed to
allow the user to check for errors in the assembly phase.
The START instruction of statement 30 indicates that specification of the model is complete and execution should pro-
ceed. It serves, also, to specify the duration of the simulation run. This is done in terms of the total number of
transactions which are to pass through the system to the
TERMINATE block(s) during the run - *not* by specifying the
total run-time as in a continuous formulation. A total 'termination count', comprising the transactions entering all
TERMINATE blocks, is recorded by the compiler and the run
is concluded when this equals the number specified in subfield A of the START command - 1500 in this case. In this
example, advantage has been taken of the additional option
to have interim results of the simulation printed out at
'snap intervals' during the run, after each set of 100
transactions has been terminated within the total of 1500
(subfield C). This is a helpful option in 'debugging' a
model and for initial examination of the results.

5..3..3. Output of the model
Selected items from the output of the example are reproduced
in Figs. 5.2 and 5.3. In Fig.5.2 we show the block counts
at snap intervals 1 of 15, 10 of 15, and 15 of 15, i.e. after
100, 1000, and 1500 terminations of transactions. Fig.5.3
shows the frequency classes of females in the breeding herd
according to their numbers of breeding cycles, at the same
intervals in the simulation.

At snap 1, after 100 animals have passed through the
system in 1911 days, the priming of the system by the GENERATE
block is complete and an additional 51 animals have been born
within the herd. Sixty males and 33 young females have been
sold. No females have yet been culled and the corresponding

(a)

THIS IS SNAP   1 OF   15

RELATIVE CLOCK   1911 ABSOLUTE CLOCK   1911
BLOCK COUNTS

BLOCK	CURRENT	TOTAL	BLOCK	CURRENT	TOTAL	BLOCK	CURRENT	TOTAL
1	0	100	11	0	37	21	0	2
2	0	151	12	0	136	22	0	33
3	6	71	13	37	136	23	0	0
4	0	65	14	0	99			
5	0	65	15	0	99			
6	0	60	16	0	99			
7	8	80	17	0	102			
8	0	72	18	0	51			
9	0	72	19	0	5			
10	0	70	20	0	60			

(b)

THIS IS SNAP   10 OF   15

RELATIVE CLOCK   14328 ABSOLUTE CLOCK   14328
BLOCK COUNTS

BLOCK	CURRENT	TOTAL	BLOCK	CURRENT	TOTAL	BLOCK	CURRENT	TOTAL
1	0	100	11	0	298	21	0	32
2	0	1136	12	0	2062	22	0	219
3	22	556	13	83	1847	23	0	215
4	0	534	14	0	1764			
5	0	534	15	0	1764			
6	0	498	16	0	1764			
7	31	580	17	0	2072			
8	0	549	18	0	1036			
9	0	549	19	0	36			
10	0	517	20	0	498			

(c)

THIS IS SNAP   15 OF   15

RELATIVE CLOCK   17568 ABSOLUTE CLOCK   17568
BLOCK COUNTS

BLOCK	CURRENT	TOTAL	BLOCK	CURRENT	TOTAL	BLOCK	CURRENT	TOTAL
1	0	100	11	0	494	21	0	54
2	0	1735	12	0	3236	22	0	333
3	28	810	13	163	2905	23	0	331
4	0	782	14	0	2742			
5	0	782	15	0	2742			
6	0	722	16	0	2742			
7	44	925	17	0	3270			
8	0	881	18	0	1635			
9	0	881	19	0	60			
10	0	827	20	0	722			

Fig.5.2. Block counts, showing current contents and total number of transactions for each block after (a) 100, (b) 1000, and (c) 1500 termination
counts.

frequency table in Fig.5.3 shows that the oldest breeding
females are in their fourth cycle at this time. Five males
and two females have died in this first period.

The outputs at snap 10 and snap 15 show that the system
is tending to stabilize, with rather similar numbers of
females in the breeding cycle frequency classes. It is
apparent, however, that the sale of 40 per cent of young
females is a slightly conservative management decision, with
overall herd numbers continuing to increase at that level
of offtake.

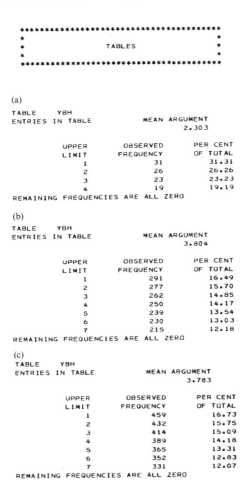

Fig.5.3. Table summaries of number of breeding females after (a) 100, (b) 1000, and (c) 1500 termination counts.

5.3.4. *Extension of the model to illustrate the queueing feature*

An important feature of discrete computer models of this type is the facility to detect any bottlenecks in the system and to analyse the 'queues' which result. To illustrate this facility we consider an extension of the example program to include the mating of the females in the herd with a limited number of males ('bulls'). It is assumed that six bulls are available and that each female must 'run with' one of these for 28 days to ensure fertilization.

The simulation of this type of situation in GPSS is accomplished by programming the passage of transactions through a multi-unit 'processor' known as a *storage*. For our purpose, we use a storage with six units to represent the bulls. This is named MATE and its capacity is defined in the following statement at the beginning of the program, prior to the sequence of blocks.

```
MATE    STORAGE    6
```

In addition, the following code is inserted between the blocks numbered 12 and 13 in the original program.

```
QUEUE       MATE,1
ENTER       MATE,1
DEPART      MATE,1
ADVANCE     28
LEAVE       MATE,1
```

The ENTER, ADVANCE, and LEAVE blocks define the processes of transactions entering the storage called MATE, occupying up to the maximum of six units representing the bulls for a period of 28 days and 'freeing' those units after that period. When insufficient spaces are available in the storage for transactions arriving at the ENTER block they are denied entry until the occupying transactions move on. The QUEUE and DEPART blocks are inserted in the program immediately before and after the ENTRY block so as to collect statistics on any queue which builds up at this point.

A portion of the output from the model with this extension is given in Fig.5.4. It shows that although the average occupancy of the storage was only 4.725 units, the bottleneck at this point resulted in a substantial queue of females waiting to be mated. Only 26 per cent of the females were put to a bull without any delay; the average number waiting was 17 and the maximum 84.

5.4. APPLICATION OF EVENT-ORIENTED MODELS

Event-oriented computer simulation so far has found fewer

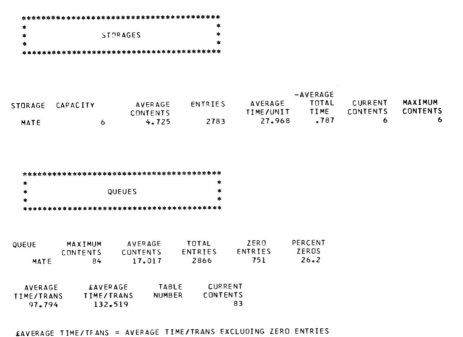

STORAGE	CAPACITY	AVERAGE CONTENTS	ENTRIES	AVERAGE TIME/UNIT	-AVERAGE TOTAL TIME	CURRENT CONTENTS	MAXIMUM CONTENTS
MATE	6	4.725	2783	27.968	.787	6	6

QUEUE	MAXIMUM CONTENTS	AVERAGE CONTENTS	TOTAL ENTRIES	ZERO ENTRIES	PERCENT ZEROS
MATE	84	17.017	2866	751	26.2

AVERAGE TIME/TRANS	£AVERAGE TIME/TRANS	TABLE NUMBER	CURRENT CONTENTS
97.794	132.519		83

£AVERAGE TIME/TRANS = AVERAGE TIME/TRANS EXCLUDING ZERO ENTRIES

Fig.5.4. The storage MATE with its associate queue of transactions at termination count 1500.

applications in biology and agriculture than continuous models. See Holling (1966) for an example of this type of modelling in an analysis of insect populations and Geisler, Paine, and Geytenbeek (1977) for a study of alternative mating policies in sheep flocks.

The ease and economy with which the dynamics of populations may be represented in event-orientated models is of particular relevance to the analysis of population behaviour under more or less stable conditions. Where there is a special concern with the precise, chronological sequence of changes, for example in the establishment of a 'new' population or in the transition period following an alteration in management policy, the serial record from a continuous formulation may be of advantage. This distinction, however, is principally one of convenience, based on the 'normal' form of output from the two model forms. The user may adapt the output of either form as required.

Within the general class of mathematical models treating events rather than continuous changes, two special forms should be mentioned here. *Markov chain* models have been employed in stochastic representations of weather variation (Feyerham and Bark 1967; Edelsten 1976) and in modelling the spread of animal diseases (James 1977). Leslie's Matrix model (Leslie 1945) and developments from it (Usher 1972) have provided an elegant and powerful tool for studies of age structures in animal populations.

6
TESTING AND USING MODELS

6.1. THE CONTEXT OF MODEL-TESTING

The primary objective in constructing a quantitative model is to *summarize* the available data and to make some form of *prediction*. This is not to underestimate the value of the 'spin-off' from an overall modelling effort, as distinct from the direct use of the model itself (Innis 1975). Neither is it intended to suggest that a model is valuable only if it serves to make 'accurate' predictions at the first attempt and is to be discarded if it does not. Such a view would imply that modelling is some special activity, divorced from the basic scientific procedure of alternately formulating hypotheses and subjecting those hypotheses to experimental and other tests which attempt to falsify them. Mechanistic models of biological and agricultural systems tend to be more complicated than the hypotheses to which many of us are accustomed - chiefly because these systems are indeed complicated objects. But systems models are certainly hypotheses and must be subjected to the usual process of attempting to disprove them if they are to make a worthwhile contribution to scientific knowledge. As Jeffers (1976) puts it, 'a model which is incapable of verification is an essay in metaphysics and not an expression of the scientific method'.

Testing or *evaluation* of the predictive power of a model is, therefore, a key element in any worthwhile modelling project. But to put the procedures for testing in their correct context it is important to recall the fundamental property of all models, that they are deliberate simplifications of reality. Consequently, we can make no sense of testing a model except in relation to its pre-defined purpose. The discussions of model testing in the literature unfortunately are not completely consistent in their use of terminology: but the association of the word *validation* with establishing a model's degree of fitness for a given purpose (Wright 1971) appears to have gained some general acceptance and is a useful reminder of this basic concept in testing.

Considering the wide range of activities in the whole spectrum of research-and-development studies in agriculture and the related biological sciences, it follows that there is a correspondingly wide range of legitimate modelling objectives which need to be taken into account in considering methods of model-testing. Two recent publications which illustrate this range of objectives and provide a good introduction to model-testing in general are Arnold and de Wit (1976) and Penning de Vries (1977). Here, we examine two markedly contrasting situations at opposite ends of the spectrum.

The first is the case of what Penning de Vries has called 'scientifically interesting' models, with the general objectives of helping to present an integrated picture of the system under study and to bridge the gap between knowledge at different degrees of resolution in the hierarchy of systems. Good examples of such models in the areas of plant physiology and soil science are de Wit, Brouwer, and Penning de Vries (1970), de Wit and van Keulen (1972), and Thornley (1976). Similarly detailed models have been developed in various areas of animal physiology, notably for the chemistry and physics of rumen function, e.g. Baldwin, Koong, and Ulyatt (1977). Typically, it is difficult to define precise performance targets for the predictions of such models, partly because of the exploratory nature of the research on which they are based and partly because the predictions of systems outputs are seldom directly applicable in agricultural practice. The ultimate evaluation of this type of modelling project depends on an assessment of the total research effort, within which the modelling activity serves to provide a framework for gathering information and as an aid in formulating and testing hypotheses.

At the other end of the research-and-development spectrum, studies in the areas of farm management and production economics are concerned principally with the prediction of system *outputs*. Consideration of the biological mechanisms which give rise to those outputs is of secondary importance except insofar as it may improve the potential for extrapolation of model predictions. The overriding concern with direct

application of the results to the management of production
enterprises makes it both natural and essential to specify
performance targets for the reliability of model predictions,
indicating the limits within which they must lie to be practical use in decision-making. Examples of modelling projects
in this applied sphere include Flinn and Musgrave (1967);
Maxwell,Eadie, and Sibbald (1973); and Edelsten and Newton
(1977).

6.2. METHODS OF TESTING
6.2.1. *Testing of outputs and components*
All mechanistic models can and should be evaluated at both
the level of overall system *outputs* and at the level of internal *components* and processes. In a model of animal production at pasture, for example, it is important to check
that major internal elements such as the amount of herbage
produced and the rate of animal intake are well predicted
by the model, in addition to the amounts of animal products
derived from the whole system. In a soil-water/crop-production
model, the prediction of water available to the crop should
be scrutinized because it is an important intermediate step
in the overall calculation of crop production. Testing at
both levels is essential because there is a real risk, otherwise, that apparently good predictions of system outputs are
derived from a model with compensating internal errors. The
frequent occurrence of negative-feedback loops in biological
systems can mask the effects of erroneous data on system components if model performance is judged only at the level of
system outputs (van Keulen 1976).

   At a superficial level, it might be argued that agree-
ment of model predictions with test data at the level of
system outputs is the principal or only criterion on which a
model should be judged in the context of farm-management operations. But this ignores the critical point that in such a
context there is no real justification for choosing a mechanistic type of model except that it may prove more reliable for
extrapolation than a simple, regression-type model. If the
model does not simulate the behaviour of the internal system
components reliably, then that possibility is excluded and the

only effective options open to the analyst are to improve
the representation of the internal components or to substitute
an input/output formulation which simply summarizes the data
on gross inputs and outputs.

6.2.2. *Qualitative and quantitative comparisons*

Prior to embarking on quantitative tests, it is good practice
to compare the predictions of a model with test data from the
real system on a 'qualitative' basis, paying particular
attention to the shapes of the time courses of both internal
variables and system outputs and to discontinuities (Penning
de Vries 1977). Such an initial evaluation should include,
if possible, a comparison of behaviour with one or more input
variables at the extremes of their possible ranges, e.g. with
very high and low stocking rates in a grazing system, so as
to test model behaviour over a wide range of system states.
Qualitative comparisons often are conveniently made in graphical form and they can serve to reveal major discrepancies
which make it superfluous to impose more precise tests until
they can be corrected.

In considering *quantitative* comparisons it is important
to establish exactly what is being compared. Test data
derived from measurements on a physical example of a system
provide an *estimate* of the behaviour of that particular example. If the ultimate objective is to characterize the behaviour of a general *class* of systems relating, for example,
to milk-production systems of the same type on a number of
individual farms, the test data should reflect the behaviour
of the class and the variability within it. Similarly, the
model predictions should be derived from a stochastic formulation for a realistic comparison of behaviour within such
a class. Meaningful comparisons and efficient use of research resources require a degree of balance between the test
data and the model predictions. If the form and reliability
of the data used in model construction allow only an exploratory, deterministic formulation, it may be more appropriate
to devote the available resources to improving the data base
for the model rather than collecting elaborate sets of test
data. Conversely, the first priority should go to obtaining

more comprehensive test data where these do not match the
form of outputs already attained by the model.

The choice of appropriate statistical procedures in
quantitative testing depends on the nature of the attributes
to be compared. Generally, the comparison of simple 'means'
presents no particular problems (Naylor, Balintfy, Burdock,
and Chu 1966; Dale 1970). Analysis of dynamic behaviour,
in terms of the 'time courses' of state variables, can be
complicated by the fact that successive values in these time
series are correlated. Quenouille (1957) and Teichroew
(1965) discuss some of the problems in detailed analyses
of this type.

6.2.3. *Internal testing*
An essential preliminary to comparative testing is to ensure
that the working model conforms exactly to what was intended.
Checking that it is free from errors in the numerical data,
that the dimensions of variables are consistent throughout
and that its detailed structure agrees with the original
specification, are difficult and time-consuming tasks in a
model of any complexity (Penning de Vries 1977; van Keulen
1976).

Some logical errors in model construction may be detected
by the compiler in special-purpose simulation languages; but
there is no systematic method of eliminating the majority of
such errors and it cannot be overemphasized that extreme
care is required.

6.3. USING MODELS
The ultimate results of a modelling project are the predictions by the model of system behaviour. The uses to which
the predictions may be put will vary according to the degree
of confidence which has been established in their reliability
and also according to the level in the hierarchy of systems
in biology and agriculture at which the model has been built.
Predictions from models of agricultural production systems
at a relatively coarse level of resolution may be used to
suggest ways of managing those systems in farming practice,
either for direct formulation of recommendations to farmers

or, more commonly, as a basis for the final stages of field
testing prior to advising farmers. The principal use of the
predictions from models incorporating the results of fundamental research at a detailed level of resolution is normally
to provide a *summary* of system behaviour for the benefit of
studies at a coarser level of understanding.

There are, however, a variety of other benefits which can
accrue from the overall modelling activity and in some circumstances these may be of greater importance than the direct
application of model predictions. They are best explained in
relation to an overall strategy for building and using models
illustrated in Fig.6.1.

6.3.1. *Models as research tools*

Individual modelling projects may not embrace all the possible
operations shown in the generalized scheme of Fig.6.1. In
the first part of the sequence, items 3 and 4 relate only to
mechanistic models. But with this exception, the operations
up to and including the testing phase of item 6 constitute
a basic, minimum subset for any modelling project concerned
with biological/agricultural systems. Having reached item
6, the modeller must decide whether it is necessary to return
to items 1 and/or 3 in order to remedy any deficiencies re-
vealed by the testing procedure - as indicated by the broken
arrows in the diagram. In other words, having formulated an
hypothesis on some aspect(s) of system behaviour, an attempt
is made to falsify that hypothesis and is followed, as necessary, by a reformulation of the hypothesis and a further
testing phase. Operating within this basic cycle, modelling
may be described as a *research tool*, with the primary function
of helping to pinpoint areas of inadequate knowledge. In exploratory research at a detailed level of understanding, this
may be the chief or only objective of a modelling project.

An essential stage in the basic cycle is to examine and
test the model predictions: otherwise there is no objective
way of assessing its deficiencies. The fact that the predictions are used in this way may be contrasted with the
direct application of predictions from models of production
systems in devising management strategies. But the primary

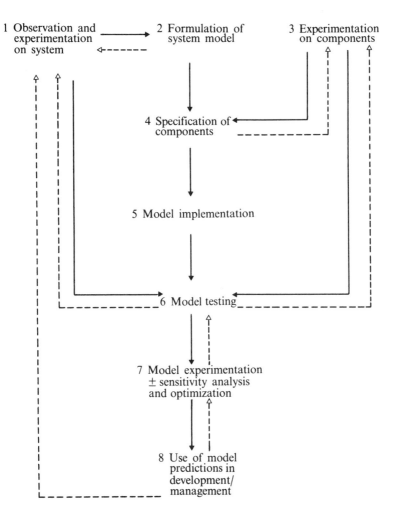

Fig.6.1. Strategy for modelling and model use.

step of making predictions is as necessary in using models
as research tools as it is in the so-called 'predictive'
models employed in management studies.

In refining or reformulating a model within the basic
cycle there is a significant variation in detailed strategy
according to whether an input/output model or a mechanistic
formulation is used. Errors or discrepancies in an input/
output model imply a need simply to re-fit the model. In
the corresponding situation where a mechanistic model gives

inadequate predictions of system outputs, adjustment or
'tuning' of one or more internal components in order to improve the output predictions is not only cumbersome, but
defeats the essential purpose of using a mechanistic model.
Once such arbitrary changes have been made to the internal
parameters or structure of a mechanistic model, its potential for extrapolation is nullified. The only valid procedure for improving a mechanistic model is to seek new and
more reliable data on suspect system components, usually involving further observation and experimentation. If that
procedure cannot be followed the only remaining practical
alternative is to abandon the mechanistic formulation and to
substitute an input/output version which can be directly
fitted.

Another aspect of using modelling as a research tool
within the basic cycle is that in some instances it may
become apparent at an early stage in model construction that
there are such major gaps in the information available that
it is impossible or unprofitable to attempt to complete the
initial assembly of a model of the whole system. This con-
tingency is accounted for in Fig.6.1 by the broken arrow
leading from item 4 to item 3. Again, it implies a necessity
to return to measurement and experimentation at the system
component level, as an essential preliminary to progressing
the overall project. It may be remarked, however, that
apart from extreme cases where the issue cannot be in doubt,
a genuine need to pursue component studies in greater detail
can only be *demonstrated* by the inadequacy of a whole-system
model based on existing data. It is sound practice, therefore, to attempt to complete the overall model at an early
stage, as a corrective to the natural tendency to become
absorbed in detailed component studies to the exclusion of
the primary objective at the whole-system level. Such an
attempt at least may provide a useful means of allocating
priorities to the investigation of a number of internal components, all of which apparently require further investigation.

6.3.2. Management applications

Items 7 and 8 in Fig.6.1 represent the direct application of predictions from models of production systems to the solution of management problems. Having established some confidence in the validity of the model predictions within the basic cycle, this last part of the sequence is concerned with harnessing the results in practice. Even within this phase, it may be necessary to return to further refinement of the model within the basic cycle, as the attempts to harness the results reveal further deficiencies or the problems to be solved are redefined.

In harnessing the results of models for management purposes two types of question commonly occur. The first concerns the *relative magnitude* of system responses to the various factors which can be manipulated by the manager and the second relates to devising good or 'optimum' *combinations* of management inputs.

If the response of a system to changes in a particular factor is large it is said to be sensitive to that factor and methods of establishing relative responses are termed *sensitivity analysis*. In very simple cases it may suffice to 'experiment' on a model and deduce sensitivity from the results of successive runs in which a parameter is progressively varied. (See example exercise 4 of Chapter 3). A complete investigation of the sensitivity of a system to a number of factors, including their combinatorial effects, demands the use of formal, systematic methods of analysis. Tomovic (1963, 1970) has proposed an approach to the analysis of dynamic models in which sensitivity is defined in terms of partial derivatives. Steinhorst, Hunt, Innis, and Haydock (1978) describe an example of sensitivity analysis on a complex ecosystem model using an analysis of variance procedure with a fractional factorial design, after Shannon (1975). Examples of formal sensitivity analyses on less complicated models are given by Smith (1970) and Miller, Weidhaas, and Hall (1973).

The utility of sensitivity analysis is not confined to management applications. It can be a useful guide to those parameters which require more exact definition in the process

of model development (Steinhorst *et al.* 1978). In the early
stages of model development, however, such indications need
to be treated with caution because their validity depends on
correct formulation of the overall model *structure*.

Selecting good or 'optimum' *combinations* of management
inputs is frequently of major importance in the application
phase of a production systems study. As with sensitivity
analysis, it may be possible to obtain some useful indications of efficient combinations of inputs from simple 'experimentation' on a simulation model. In all but very simple
cases, however, more formal methods of analysis will be required.

In some situations the problem may be formulated solely
in terms of maximizing system output from the combination of
a number of inputs. Radford (1972) describes a simple example of such an exercise, in which Powell's (1964) method of
'steepest ascent' was used in conjunction with a simulation
model to estimate the maximum harvestable yield from a grass
sward with varying times of initial harvest and a range of
intervals between subsequent harvests. But the relevance
of such solutions to management problems depends on *implicit*
restriction of the inputs to those which are practically
feasible or desirable. In the majority of practical applications it is necessary to state the constraints explicitly
and the methods of solution most commonly adopted are those
of *linear programming*.

The techniques of linear programming have found widespread applications in management science generally in the
last thirty years, including a significant role in farm
management studies (see, for example, Heady and Candler
1968 and Dent and Casey 1967). Despite the restriction of
these methods to problems which can be formulated as linear
equations, they have been used successfully for a wide
variety of management studies in agriculture, ranging from
the formulation of 'least-cost' diets for livestock to whole-
farm management strategies. The methods of solution are
essentially computer-based and suitable 'packages' are available in the software libraries of many computer installations.
Some introductory and reference texts on linear programming

and its applications are included in the bibliography at the
end of this chapter.

Apart from the restriction that the function to be maxi-
mized or minimized and the constraints on the solution must
be formulated as linear equations, linear programming methods
are concerned essentially with 'static' rather than dynamic
situations, i.e. those in which the problem does not change
with time. The management of some agricultural production
systems over a period of time may involve a series of decisions which, ideally, should be considered as a set, for
the derivation of an optimal overall strategy. Optimization
problems of this type are the province of *dynamic programming*
methods, another of the types within the general class of
mathematical programming methods to which linear programming
belongs. Amidon and Akin (1968) report one of the few applications of dynamic programming to biological problems.
Swartzman and Van Dyne (1972) and Crabtree (1972) give
examples of the iterative use of linear programming routines
coupled to simulation models as an alternative approach to
this type of problem. In their approach, the linear programming routine is invoked at intervals throughout a run
of the simulation model, using information on the current
state of the system from the simulation to devise an optimum
plan for management over the interval until the next invocation. Such a series of independent optimizations may well
depart from an optimum *overall* plan, but it has the interesting analogy to practical operation of a production system
that the periodic management decisions are based only on
currently available information. Jacobs (1967) provides a
simplified, introductory treatment of dynamic programming
proper; more advanced texts are Bellman (1957), the pioneer
in this field, and Kaufman and Cruon (1967).

APPENDIX 1

SOLUTIONS TO EXAMPLE EXERCISES IN CHAPTERS 2 AND 3

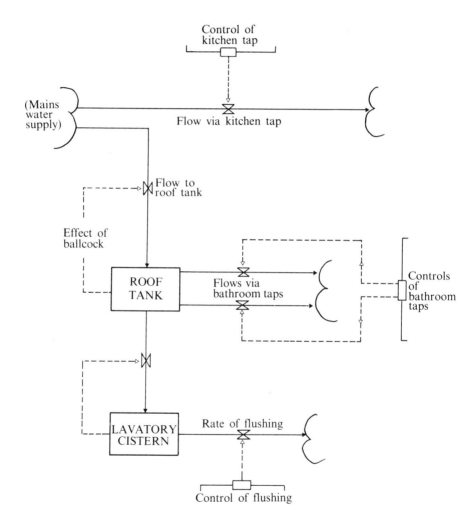

Fig.A1.1.Solution to exercise 1, Chapter 2. In comparing his own solution to the problem with the version given here the reader should check that he has included the same model components and that they are linked together in the same fashion. Within those constraints, the diagram may be drawn in a number of alternative ways which are equally acceptable. But what- ever the precise layout chosen it is worthwhile ensuring that it is neatly done, so that the salient points may be easily appreciated by others.

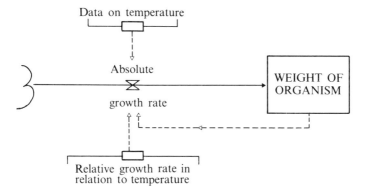

Fig.A1.2. Solution to exercise 2, Chapter 2.

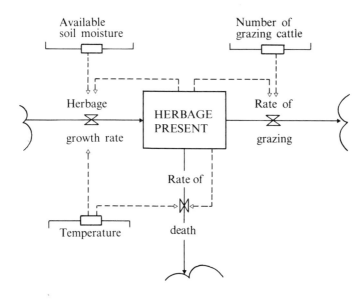

Fig.A1.3. Solution to exercise 3, Chapter 2.

```
*        CSMP PROGRAM FOR EXAMPLE EXERCISE (1)
*        OF CHAPTER III -- BACTERIAL GROWTH
INITIAL                                                          001
PARAMETER INNB=60.,PRIN=0.02                                     002
DYNAMIC                                                          003
*            FLOWS
     INCR=NB*PRIN                                                004
*            COMPARTMENTS
     NB=INTGRL(INNB,INCR)                                        005
METHOD RECT                                                      006
TIMER DELT=0.5,FINTIM=24.,PRDEL=0.5                              007
TITLE     BACTERIAL GROWTH                                       008
PRINT NB                                                         009
END                                                             010
STOP                                                            011


*        CSMP PROGRAM FOR EXAMPLE EXERCISE (2)
*        OF CHAPTER III -- PASTURE PRODUCTION
INITIAL                                                          001
PARAMETER INWOH=1200.                                            002
DYNAMIC                                                          003
     TEMP=AFGEN(TTEMP,TIME)                                      004
FUNCTION TTEMP=(0.,6.),(1.,2.),(2.,8.),(3.,24.),...             005
               (4.,29.),(5.,15.),(6.,11.),...                    006
               (7.,28.),(8.,33.),(9.,7.)                         007
     GRH=AFGEN(TGRH,TEMP)                                        008
FUNCTION TGRH=(0.,0.),(4.,0.),(25.,150.),(35.,0.)
     WOH=INTGRL(INWOH,GRH)                                       010
METHOD RECT                                                     011
TIMER DELT=1.,FINTIM=9.,PRDEL=1.                                 012
TITLE PASTURE PRODUCTION                                         013
PRINT GRH,WOH                                                    014
END                                                             015
STOP                                                            016


*        CSMP PROGRAM FOR EXAMPLE EXERCISE (3)
*        OF CHAPTER III -- PASTURE PRODUCTION
*        AND GRAZING
INITIAL                                                          001
PARAMETER INWOH=1200.                                            002
DYNAMIC                                                          003
     TEMP=AFGEN(TTEMP,TIME)                                      004
FUNCTION TTEMP=(0.,6.),(1.,2.),(2.,8.),(3.,24.),...             005
               (4.,29.),(5.,15.),(6.,11.),...                    006
               (7.,28.),(8.,33.),(9.,7.)                         007
     GRH=AFGEN(TGRH,TEMP)                                        008
FUNCTION TGRH=(0.,0.),(4.,0.),(25.,150.),(35.,0.)               009
     CONSH=AFGEN(TCONSH,WOH)                                     010
FUNCTION TCONSH=(0.,0.),(700.,0.),(1500.,120.),...              011
               (5000.,120.)                                      012
     WOH=INTGRL(INWOH,(GRH-CONSH))                               013
METHOD RECT                                                     014
TIMER DELT=1.,FINTIM=9.,PRDEL=1.                                 015
TITLE PASTURE PRODUCTION AND GRAZING                             016
PRINT GRH,CONSH,WOH                                              017
END                                                             018
STOP                                                            019
```

Fig.A1.4. Solutions to exercises 1-3, Chapter 3.

Solution to exercise 4, Chapter 3

The program may be modified so as to examine the required range of selling percentages by inserting additional PARAMETER specifications between the END and STOP instructions, thus:

```
INITIAL
- - - - - -
PARAMETER PSLF=0.26
- - - - -
DYNAMIC
- - - - - -


- - - - - -
END
PARAMETER PSLF=0.29
END
PARAMETER PSLF=0.32
END
STOP
```

The above modification will initiate three 'runs' of the program, with varying numerical values for PSLF.

Alternatively, a multiple-value parameter instruction
may be used in the INITIAL section, with the alternative
numerical values in parentheses:

```
PARAMETER PSLF=(0.25,0.29,0.32)
```

This latter form of instruction also causes three runs of the
model to be executed; but it differs from the first in that
it allows data from all three runs to appear on a single
OUTPUT document. See CSMP users' manual for details of
these alternatives.

COMPLETE FORTRAN PROGRAMS FOR THE EXAMPLE MODELS
DESCRIBED IN CHAPTER 3

The FORTRAN programs given in Chapter 3 are concerned only
with the 'core' calculations of defining the rate processes
and of allowing for the updating of the state variables.
Additional instructions are required to convert them into
complete, working versions, chiefly to permit the necessary
inputs and outputs of numerical data. Generally, the technical aspects of these operations are more complicated in
FORTRAN than in languages such as CSMP. In order to minimize
the attention which must be given to these technicalities
when modelling in FORTRAN, the complete programs described
here have been designed to make use of standard subprograms
(subroutines and functions) as far as possible.

Variations between computer installations make it impossible to design FORTRAN programs and subprograms for
universal application, without some minor changes to suit
particular machines and their associated 'software'. We
have attempted to keep these 'machine-dependent' items to a
minimum, so that the user should have little difficulty in
implementing programs on most modern computers. Nevertheless,
any doubts or problems in implementation should be referred
to the programme advisory staff at the installation concerned.

RELATIVE GROWTH MODEL
The complete listing of the program given below includes the
main program and the associated subroutines as implemented on
an ICL 470 computer.

Comparing the complete listing of the main program with
the core part given on p.46, it will be seen that the full
version is built around the time-loop containing the iterative calculations for the rates and state variables. Taking
advantage of the automatic facility for consecutive number-
ing of the lines of the program for reference purposes, these
reference numbers appear at the left-hand side of the computer printout reproduced below. The core instructions discussed in Chapter 3 correspond to lines 51 and 52, 55-7, and

```
 1     C-----------------------------------------------------------
 2     C    N R BROCKINGTON
 3     C    RELATIVE GROWTH MODEL
 4     C    STARTING WEIGHT 20
 5     C    RELATIVE GROWTH RATE 0.01 PER DAY
 6     C    CALCULATE GROWTH OVER 50 DAYS
 7     C    CALCULATION INTERVAL OF ONE DAY
 8     C
 9     C    V(8) IS THE ARRAY FOR VARIABLES
10     C
11     C    PRINTING AND PLOTTING DEVICES,ARRAYS AND HEADINGS
12     C
13     C    TITLE(60) IS THE HEADING FOR PRINTOUTS,
14     C    CODE THE CHARACTERS IN BLOCKS OF ONE AND FILL UP
15     C    THE REMAINING BLOCKS WITH BLANKS
16     C
17     C    VNAME(48) IS THE SET OF LABELS FOR OUTPUT VARIABLES
18     C    CODE THE CHARACTERS IN BLOCKS OF ONE,AND
19     C    EACH VARIABLE CAN HAVE UP TO SIX CHARACTERS IN ITS
20     C    LABEL - FILL ALL  REMAINING BLOCKS WITH BLANKS
21     C
22     C    INDEX(5) CONTAINS INDICES FOR OUTPUT VARIABLES TO
23     C    BE PLOTTED--USE ZEROS TO FILL UP THE ARRAY
24     C
25     C    COMPUTER SYSTEM CONSTANTS
26     C
27     C    MOUTV IS INTEGER NO. OF OUTPUT DATA SET REF. NO.
28     C    MINV IS INTEGER NO. OF INPUT DATA SET REF. NO.
29     C    LINEPP IS NUMBER OF LINES ON PRINTOUT PAGE
30     C    NCOLS IS NUMBER OF COLUMNS ON OUTPUT DEVICE
31     C    BIGPOS IS LARGEST REAL POSITIVE NUMBER
32     C    BIGNEG IS LARGEST REAL NEGATIVE NUMBER
33     C-----------------------------------------------------------
34           COMMON/SYSCON/MOUTV,MINV,LINEPP,NCOLS,BIGPOS,BIGNEG
35           DIMENSION TITLE(60),VNAME(48),V(8),INDEX(5)
36           DATA TITLE/'R','E','L','A','T','I','V','E',' ',
37          +'G','R','O','W','T','H',' ','R','A','T','E',
38          +' ','M','O','D','E','L',34*' '/
39           DATA VNAME/'A','G','R',' ',' ',' ',' ','W','T',' ',' ',' ',
40          +' ',' ',36*' '/
41           DATA INDEX/2,0,0,0,0/
42     C    SET VARIABLES FOR COMPUTER SYSTEM
43           MOUTV=6
44           MINV=5
45           LINEPP=45
46           NCOLS=120
47           BIGPOS=1.0E75
48           BIGNEG=-1.0E75
49           N=0
50     C    INITIALISATION AND DATA INPUTS
51           WT=20.0
52           RGR=0.01
53     C    TIME LOOP SET TO NUMBER OF CALCULATION INTERVALS
54     C    IN TOTAL RUN TIME
55           DO 100 ITIME=1,50
56           AGR=WT*RGR
57           WT=WT+AGR
58     C    STORE OUTPUT VARIABLES IN V(8)
59           V(1)=AGR
60           V(2)=WT
61     C    PRINTING OUTPUT VARIABLES
62           CALL PRINT(V,ITIME,TITLE,VNAME,2)
63     C    WRITING VARIABLES FOR GRAPH TO TEMPORARY STORE
64           CALL PLT(V,ITIME,TITLE,VNAME,INDEX)
65     C    END OF TIME LOOP
66       100 CONTINUE
67     C    PRINTING GRAPH FROM TEMPORARY STORE
68           CALL PLEND
69           STOP
70           END
```

66 in the full listing.

The additional instructions in the main program centre around three subroutines used to produce the results of running the model in tabular and graphical form. The subroutine PRINT is called in line 62 to effect a tabular listing of the chosen

```
C-----------------------------------------------------------
C     PRINT PRINTS THE VARIABLES IN V
C     WHERE NV =NO. OF VARIABLES
C-----------------------------------------------------------
      SUBROUTINE PRINT(V, ITIME, TITLE, VNAME, NV)
      COMMON/SYSCON/MOUTV, MINV, LINEPP, NCOLS, BIGPOS, BIGNEG
      DIMENSION V(8), TITLE(60), VNAME(48)
      LOGICAL BEGIN
      DATA BEGIN/. FALSE. /
      IF(BEGIN)GO TO 6
      BEGIN=. TRUE.
      L=LINEPP+2
    6 L=L+1
      IF(L. LT. LINEPP+3)GO TO 10
      L=2
      K=6*NV
      WRITE(MOUTV, 5)TITLE, (VNAME(I), I=1, K)
    5 FORMAT(1H1, 60A1/'   TIME   ', 8(6X, 6A1))
   10 WRITE(MOUTV, 15)ITIME, (V(I), I=1, NV)
   15 FORMAT(1X, 3X, I3, 3X, 8G12. 5)
      RETURN
      END


C-----------------------------------------------------------
C     PLT LOADS VARIABLES TO BE PLOTTED INTO
C     ARRAY DUMMY
C-----------------------------------------------------------
      SUBROUTINE PLT(V, ITIME, TITLE, VNAME, INDEX)
      DIMENSION V(8), TITLE(60), VNAME(48), INDEX(5), XMIN(5),
     +XMAX(5), DUMMY(50, 5), VNAME1(30), MTEMP(50)
      LOGICAL START
      COMMON/XXXXXX/TITLE, VNAME1, XMIN, XMAX, N, NV, MTEMP, DUMMY
      COMMON/SYSCON/MOUTV, MINV, LINEPP, NCOLS, BIGPOS, BIGNEG
      DATA START/. FALSE. /
      IF(START)GO TO 10
      START=. TRUE.
      N=0
      NV=0
      DO 5 I=1, 5
      K=INDEX(I)
      IF(K. LT. 1. OR. K. GT. 8)GO TO 10
      NV =NV+1
      XMIN(I)=BIGPOS
      XMAX(I)=BIGNEG
      DO 6 J=1, 6
      IJL=6*(I-1)+J
      KJL=6*(K-1)+J
    6 VNAME1(IJL)=VNAME(KJL)
    5 CONTINUE
   10 IF(NV. EQ. 0)RETURN
      N=N+1
      MTEMP(N)=ITIME
      DO 15 I=1, NV
      IK=INDEX(I)
      DUMMY(N, I)=V(IK)
      IF(V(IK). LT. XMIN(I))XMIN(I)=V(IK)
   15 IF(V(IK). GT. XMAX(I))XMAX(I)=V(IK)
      RETURN
      END
```

output variables against time. The call to PLT in line 64
writes the values of variables chosen to appear in a graph
to a temporary store. The subroutine PLEND is also concerned
with the production of the graph. It is called outside the
time loop, when the model calculations have been completed,
so that the full numerical ranges of variables to be graphed
may be examined and appropriately scaled. Having selected
appropriate scales, PLEND produces the graph on the line
printer, with the selected variables plotted against time.

Apart from the calls to the subroutines, additional

```
C---------------------------------------------------------
C      PLEND READS DATA FROM ARRAY DUMMY AND PRODUCES
C      A GRAPH
C      NB.  NCOLS=NO. OF COLUMNS ON OUTPUT DEVICE
C---------------------------------------------------------
       SUBROUTINE PLEND
       DIMENSION TITLE(60),VNAME(30),XMIN(5),XMAX(5),
      +SYMBOL(5),DUMMY(50,5),OUT(117),MTEMP(50)
       COMMON/XXXXXX/TITLE,VNAME,XMIN,XMAX,N,NV,MTEMP,DUMMY
       COMMON/SYSCON/MOUTV,MINV,LINEPP,NCOLS,BIGPOS,BIGNEG
       DATA SYMBOL,BLANK,BAR,LINE/'A','B','C','D','E',' ',
      +'!','-'/
       IF(NV.EQ.0)RETURN
       WRITE(MOUTV,21)TITLE
   21  FORMAT('1',60A1//'VARIABLE SYMBOL  MINIMUM
      +'MAXIMUM')
       DO 25 K=1,NV
       KB=6*(K-1)+1
       KE=6*(K-1)+6
   25  WRITE(MOUTV,26)(VNAME(KV),KV=KB,KE),
      +SYMBOL(K),XMIN(K),XMAX(K)
   26  FORMAT(1X,6A1,5X,A1,3X,2G13.5)
       IN=NCOLS-13
       WRITE(MOUTV,42)(LINE,I=1,IN)
       NOUT=NCOLS-15
       DO 28 I=1,NV
       XMAX(I)=XMAX(I)-XMIN(I)
       IF(XMAX(I).LE.0.0)XMAX(I)=1.0
   28  XMAX(I)=NOUT/XMAX(I)
       DO 50 J=1,N
       DO 35 I=1,NOUT
   35  OUT(I)=BLANK
       DO 40 I=1,NV
       K=(DUMMY(J,I)-XMIN(I))*XMAX(I)+1.5
       IF(K.GT.NOUT)K=NOUT
   40  OUT(K)=SYMBOL(I)
       WRITE(MOUTV,41)MTEMP(J),BAR,(OUT(K),K=1,NOUT),BAR
   41  FORMAT(1X,G12.5,1X,119A1)
   50  CONTINUE
       WRITE(MOUTV,42)(LINE,I=1,IN)
   42  FORMAT(14X,119A1)
       RETURN
       END
```

instructions in the main program are required to give effect
to those calls, to specify which variables are to appear in
the output and how they are to be labelled.

Outline instructions for the use of the output devices
are included on the comment cards in the listing of the program; but we examine, below, the purposes for which they are
designed and how they are used in this particular model.

Lines 34 and 35 are standard instructions to establish
the necessary common blocks and to dimension the arrays. They
should not be varied by the user unless it is required to
modify the way in which the subroutines operate.

The DATA statement in lines 36-8 allows the user to
insert a title of his choice which will appear at the head of
both the tabular and graphical output. Allowance is made for
up to 60 characters and spaces in the title; the remainder of
the array must be filled with blanks. In this example, 34
blanks remained to be filled after inserting the title

'RELATIVE GROWTH RATE MODEL'.

   The following DATA statement, in lines 39-40, specifies
the names which will identify the output variables in the
results. Provision is made up for up to eight such identifying labels, each with a maximum of six characters. Two labels
are used in this example, AGR and WT. Notice, again, how
blanks must be specified to fill unused spaces, both within
the variable labels themselves if they do not use the full
six characters and within the whole array.

   The third DATA statement, in line 41, specifies which
of the variables appearing in the tabular output are also to
be graphed. The reference to the variables to be plotted in
the graph is by the numbers in an array, V, the subscripts
of which are the numbers 1 to 8. All the variables to appear
in tabular and/or graphical form are stored in this array
and in this example the variables AGR and WT are stored in
$V(1)$ and $V(2)$ in lines 59 and 60. Since a graph of WT is
required, the number 2 appears in the DATA INDEX statement,
followed by four zeros. As is implied by the sizes of the
V and INDEX arrays, provision is made in the design of the
output devices for up to eight variables to appear in the
tabular output and for up to five of these to be plotted in
the line-printer graph.

   Lines 43-8 are to provide details of the particular computer installation so that the subroutines will be effective
on a given machine. Once established, they may be used as a
standard block of coding on an individual machine, of course.
Lines 43 and 44 provide the integer code numbers for the
writing and reading devices, 6 and 5 in this example, which
appear on the FORMAT statements accompanying WRITE and READ
commands. LINEPP gives the number of lines per page on the
line printer and NCOLS the number of characters per line.
BIGPOS and BIGNEG are the largest real positive and negative
numbers which can be held within the machine, used in the
programming of the PLT subroutine.

   Setting the variable N to zero in line 49 is a standard
instruction required for the calculations in the subroutines.

   As we noted earlier, lines 51 and 52, 55-7 and line 66
are identical with those described in Chapter 3.

Of the additional instructions within the time loop, lines 59 and 60 store the variables chosen for output in the array, V, which is one of the 'arguments' for the subroutines PRINT and PLT. In calling PRINT, in line 62, the user must set the final argument within the parentheses to the number of variables to be tabulated, two in this case. The remaining arguments in calling PRINT and all those for calling PLT are standard for the user of these devices.

Having terminated the time loop in line 66, the call to PLEND in line 68 completes the instructions for obtaining the graphical output.

The subroutines PRINT, PLT, and PLEND are listed without comment, being designed to effect a standard form of output for the user who does not wish to be involved in the technicalities of that operation. Experienced FORTRAN programmers may choose to modify or replace them with instructions which are specifically designed for their particular needs. There is scope, of course, for modifications and additions which would provide some of the flexibility and sophistication of the output facilities in languages such as CSMP. But the relatively wide availability of the specially designed simulation languages and the considerable investment of time and professional expertise which have been involved in their evolution make it unprofitable to attempt to duplicate such facilities in a comprehensive fashion.

The tabulated and graphical outputs from the relative growth model in the FORTRAN version are shown in Figs. A2.1 and A2.2. Comparison with the output from the CSMP version (Figs. 3.2 and 3.3; pp. 44 and 45) shows the basic similarity of the form of the output in this simple example, without recourse to the sophisticated features of CSMP.

However, there is one major technical difference which needs to be mentioned. This concerns the time of printing and plotting the numerical values for the output variables in relation to the iterative calculations for the rates and state variables. In CSMP, the effective sequence of operations at a given point in time involves updating the state variables, calculating the rate variables as projected values over the *next* time-step and then printing out those values.

RELATIVE GROWTH RATE MODEL

TIME	AGR	WT
1	0.20000	20.200
2	0.20200	20.402
3	0.20402	20.606
4	0.20606	20.812
5	0.20812	21.020
6	0.21020	21.230
7	0.21230	21.443
8	0.21443	21.657
9	0.21657	21.874
10	0.21874	22.092
11	0.22092	22.313
12	0.22313	22.536
13	0.22536	22.762
14	0.22762	22.989
15	0.22989	23.219
16	0.23219	23.451
17	0.23451	23.686
18	0.23686	23.923
19	0.23923	24.162
20	0.24162	24.404
21	0.24404	24.648
22	0.24648	24.894
23	0.24894	25.143
24	0.25143	25.395
25	0.25395	25.648
26	0.25648	25.905
27	0.25905	26.164
28	0.26164	26.426
29	0.26426	26.690
30	0.26690	26.957
31	0.26957	27.226
32	0.27226	27.499
33	0.27499	27.774
34	0.27774	28.051
35	0.28051	28.332
36	0.28332	28.615
37	0.28615	28.901
38	0.28901	29.190
39	0.29190	29.482
40	0.29482	29.777
41	0.29777	30.075
42	0.30075	30.375
43	0.30375	30.679
44	0.30679	30.986
45	0.30986	31.296
46	0.31296	31.609
47	0.31609	31.925
48	0.31925	32.244
49	0.32244	32.567
50	0.32567	32.892

Fig.A2.1. Tabular output from the relative growth model.

At time = 10, for example, the state variables will be updated
using the projections for the rates derived from the *previous*
iteration, the rate calculations will be estimated for the
next time-step and both state variables and projected rates
will be printed with time recorded as the value 10. In the
FORTRAN version the sequence at each iteration involves projecting the rates over the next time-step, followed by updating the state variables and the printing of the results.

   The result of printing the output variables at a different
point in the calculation sequence may be seen by comparing
the values for the rate variable, AGR, in Figs.3.2 and A2.1.

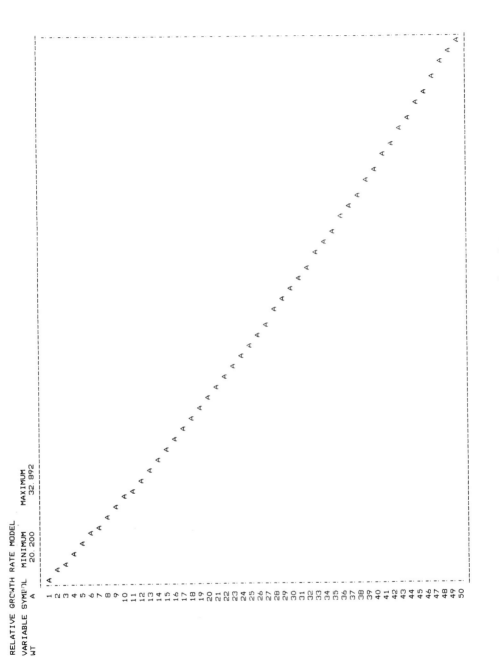

RELATIVE GROWTH RATE MODEL

VARIABLE	SYMBOL	MINIMUM	MAXIMUM
WT	A	20.200	32.892

Fig.A2.2. Graphical output from the relative growth model.

Throughout the simulation run, the value for this variable
in the CSMP results relates to the *following* time-step,
whereas in the FORTRAN version it relates to the *previous* time-
step. The values for the state variable, WT, coincide in the
two versions of the results because they are related, in
both sequences, to a given *point* in time rather than to a
time *interval*, the calculation interval.

This technical difference generally is of little practical significance, because most applications involve the use
of a calculation interval which is much smaller in relation
to the overall simulation period than we have used in this
simple example. But it can be of some significance where
the time-step between successive iterations is relatively
large. In such circumstances it may be worthwhile elaborating
the structure of a FORTRAN simulation program to correct, also,
the minor anomaly that the results do not include the values
of the output variables at time zero. In the interests of
simplicity, the structure we have adopted in these example
progams includes only one set of output instructions and these
commands follow the calculations of the rate processes and
the updating of the state variables within the time loop.
This structure makes no provision for the output of results
at time zero because by the stage when the output instructions are encountered the state variables already have been
updated from their initial values. Output of results at
time zero may be achieved by adopting the following revised
structure, which also allows for the output to occur at the
same time within the calculation sequence as in CSMP, i.e.
with values for the rate processes relating to the *following*
time-step rather than the *previous* one.

(a)   Prior to entering the time loop
   (i)   Initialization of state variables and other
            data inputs.
   (ii)   Calculation of rates.
   (iii)   Calls to PRINT and PLT, with associated
            output instructions.

(b)   Within the time loop
   (i)   Updating of state variables
   (ii)   Calculation of rates
   (iii)   Calls to PRINT and PLT.

Notice the requirement to evaluate the rate equations before entering the time loop in this revised structure and the associated reversal of the order in which the rates and state variables are calculated within the time loop.

Before leaving the topic of the timing of the output instructions within the calculation sequence, it may be noted that while the effect of the procedures in CSMP is as described above when the rectangular method of integration is chosen, the situation is necessarily more complicated with other integration methods. The CSMP users' manual provides details of how these other methods are accommodated, using a feature known as 'centralized integration'.

CARBON METABOLISM MODEL

The computer listing below includes the main program and the function TABFN, which is used to provide linear interpolation between a series of co-ordinates in the definition of the unit rate of photosynthesis. The three subroutines to effect the tabular and graphical output of results are not repeated, as they are identical with those used in the relative growth model.

The structure of the main program for the carbon metabolism model is exactly analogous to that of the relative growth model, using the same definitions of machine-dependent parameters and the same devices for the output of the results.

Two minor variations in the use of the output mechanisms relate to the frequency of printing and plotting the output variables and the number of variables to appear on the graph. The instruction in line 72 bypasses the output instructions except when the current index of the time loop is exactly divisible by two, so that the results are printed out at every second iteration of the loop (Figs. A2.3 and A2.4). Inclusion of different divisors within this form of instruction

```
 1    C----------------------------------------------------------------
 2    C   N R BROCKINGTON
 3    C   MODEL OF CARBON METABOLISM IN A GREEN PLANT
 4    C
 5    C   V(8) IS THE ARRAY FOR VARIABLES
 6    C
 7    C   PRINTING AND PLOTTING DEVICES, ARRAYS AND HEADINGS
 8    C
 9    C   TITLE(60) IS THE HEADING FOR PRINTOUTS,
10    C   CODE THE CHARACTERS IN BLOCKS OF ONE AND FILL UP
11    C   THE REMAINING BLOCKS WITH BLANKS
12    C
13    C   VNAME(48) IS THE SET OF LABELS FOR OUTPUT VARIABLES
14    C   CODE THE CHARACTERS IN BLOCKS OF ONE, AND
15    C   EACH VARIABLE CAN HAVE UP TO SIX CHARACTERS IN ITS
16    C   LABEL - FILL ALL REMAINING BLOCKS WITH BLANKS
17    C
18    C   INDEX(5) CONTAINS INDICES FOR OUTPUT VARIABLES TO
19    C   BE PLOTTED -- USE ZEROS TO FILL UP THE ARRAY
20    C
21    C   COMPUTER SYSTEM CONSTANTS
22    C
23    C   MOUTV IS INTEGER NO. OF OUTPUT DATA SET REF. NO.
24    C   MINV IS INTEGER NO. OF INPUT DATA SET REF. NO.
25    C   LINEPP IS NUMBER OF LINES ON PRINTOUT PAGE
26    C   NCOLS IS NUMBER OF COLUMNS ON OUTPUT DEVICE
27    C   BIGPOS IS LARGEST REAL POSITIVE NUMBER
28    C   BIGNEG IS LARGEST REAL NEGATIVE NUMBER
29    C----------------------------------------------------------------
30          COMMON/SYSCON/MOUTV,MINV,LINEPP,NCOLS,BIGPOS,BIGNEG
31          DIMENSION TITLE(60),VNAME(48),V(8),INDEX(5)
32          REAL NPO
33          REAL TRATE(8)
34          DATA TRATE/1.,0.10,11.,0.075,21.,0.05,91.,0.04/
35          DATA TITLE/'C','A','R','B','O','N',' ',
36         +'M','E','T','A','B','O','L','I','S','M',
37         +' ','M','O','D','E','L',37*' '/
38          DATA VNAME/'P','O',' ',' ',' ',' ',' ','N','P','O',
39         +' ',' ',' ',' ',36*' '/
40          DATA INDEX/1,2,0,0,0/
41    C   SET VARIABLES FOR COMPUTER SYSTEM
42          MOUTV=6
43          MINV=5
44          LINEPP=48
45          NCOLS=120
46          BIGPOS=1.0E75
47          RIGNEG=-1.0E75
48          N=0
49    C   INITIALISATION AND DATA INPUTS
50          CHP=0.0
51          PO=300.0
52          NPO=200.0
53          PRT1=0.7
54          PRT2=0.3
55          PRR1=0.015
56          PRR2=0.020
57    C   TIME LOOP SET TO NUMBER OF CALCULATION INTERVALS
58    C   IN TOTAL RUN TIME
59          DO 300 ITIME=1,90
60    C   FLOWS
61          TIME=FLOAT(ITIME)
62          URATE=TABFN(TRATE,4,TIME)
63          RPHOT=PO*URATE
64          RT1=CHP*PRT1
65          RT2=CHP*PRT2
66          RR1=PO*PRR1
```

serves to output the results at any desired multiple of the calculation interval. Two variables are requested in the graph in the instruction in line 40. It will be noticed in comparing Fig.A2.4 with Fig.3.6 (p.58) that the variables NPO and PO are plotted on independent scales in the scales in the graph from the FORTRAN version. The subroutine PLEND does not allow for plotting a number of variables on a common scale - as may be done in CSMP with the GROUP

```
67              RR2=NPO*PRR2
68      C  COMPARTMENTS
69              CHP=CHP+RPHOT-RT1-RT2
70              PO=PO+RT1-RR1
71              NPO=NPO+RT2-RR2
72              IF (MOD(ITIME,2).GT.0)GO TO 300
73      C  STORE OUTPUT VARIABLES IN V(8)
74              V(1)=PO
75              V(2)=NPO
76      C  PRINT OUTPUT VARIABLES
77              CALL PRINT (V,ITIME,TITLE,VNAME,2)
78      C  WRITE VARIABLES FOR GRAPH TO TEMPORARY STORE
79              CALL PLT(V,ITIME,TITLE,VNAME,INDEX)
80      C  END OF TIME LOOP
81          300 CONTINUE
82      C  PRINTING GRAPH FROM TEMPORARY STORE
83              CALL PLEND
84              STOP
85              END
```

```
C--------------------------------------------------------
C      TABFN INTERPOLATES IN A TABLE. TABLE IS IN FORM
C      X1,Y1,X2,Y2,ETC., WHERE X IS MONOTONICALLY INCREASING
C      WHERE NTABLE IS THE NO. OF PAIRS OF VARIABLES
C      TABFN IS RETURNED AS THE VALUE OF Y CORRESPONDING
C      TO THE VALUE OF X.
C--------------------------------------------------------
       FUNCTION TABFN(TABLE,NTABLE,X)
       DIMENSION TABLE(100)
       J=NTABLE*2-1
       DO 10 I=3,J,2
       IF(X-TABLE(I).LT.0.0)GO TO 20
    10 CONTINUE
       I=J
    20 XO=TABLE(I)
       YO=TABLE(I+1)
       TABFN=YO+(X-XO)*(TABLE(I-1)-YO)/(TABLE(I-2)-XO)
       RETURN
       END
C--------------------------------------------------------
```

instruction.

The specification of URATE using the TABFN function in line 62 is explained in Chapter 3. To implement this specification in the working program requires that the array TRATE be dimensioned, as in line 33, with eight elements to accommodate the four pairs of numerical values. In line 34 the numerical data are specified, each value for the independent variable being followed by the corresponding value for the dependent variable. Notice that the independent variable, TIME, is specified as a 'real' variable in line 61 to avoid the technical anomaly of mixed real and integer values in the array TRATE. In addition, the values for TIME in line 34 range from 1.0 to 91.0, rather than from 0.0 to 90.0 as in the CSMP version. Again, this is related to the point in the calculation sequence at which the output of the results occurs in the FORTRAN and CSMP versions and the alternative specification ensures the exact coincidence of the numerical values for the state variables.

```
              CARBON METABOLISM MODEL
              TIME        PO          NPO
                  2      312.07       201.08
                  4      343.39       210.48
                  6      375.69       220.33
                  8      408.55       230.43
                 10      441.56       240.62
                 12      474.31       250.71
                 14      506.32       260.50
                 16      537.11       269.79
                 18      566.17       278.37
                 20      592.99       286.01
                 22      617.07       292.53
                 24      640.99       299.03
                 26      665.58       305.85
                 28      690.85       313.02
                 30      716.82       320.52
                 32      743.48       328.35
                 34      770.85       336.51
                 36      798.93       345.00
                 38      827.73       353.82
                 40      857.23       362.96
                 42      887.46       372.42
                 44      918.41       382.20
                 46      950.09       392.29
                 48      982.49       402.70
                 50      1015.6       413.42
                 52      1049.5       424.44
                 54      1084.0       435.77
                 56      1119.3       447.40
                 58      1155.3       459.32
                 60      1192.0       471.52
                 62      1229.4       484.02
                 64      1267.5       496.79
                 66      1306.3       509.84
                 68      1345.8       523.15
                 70      1385.9       536.72
                 72      1426.7       550.56
                 74      1468.2       564.63
                 76      1510.3       578.95
                 78      1553.0       593.51
                 80      1596.3       608.29
                 82      1640.2       623.28
                 84      1684.7       638.49
                 86      1729.7       653.89
                 88      1775.2       669.48
                 90      1821.3       685.25
```

Fig.A2.3. Tabular output from the carbon metabolism model.

ANIMAL POPULATION MODEL

The computer listing of the FORTRAN version of this model
given below contains the main program and a function for
generating pseudo-random numbers, RNDX. The subroutines
to generate tabular and graphical output and the function
TABFN are required to run this model but are not repeated
in the listing because they are identical with the forms
in the previous examples.

Minor variations in usage in the program for the
animal population model compared with the previous examples
include the specification of six variables to appear in the
tabulated output and the suppression of the graphical output
by omitting the DATA INDEX statement and the calls to PLT
and PLEND.

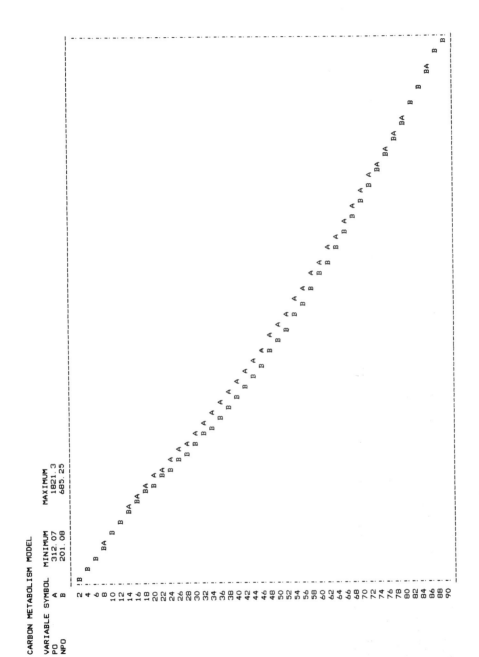

Fig.A2.4. Graphical output from the carbon metabolism model.

```
 1    C------------------------------------------------------------
 2    C    N R BROCKINGTON
 3    C    ANIMAL POPULATION MODEL
 4    C
 5    C
 6    C    PRINTING AND PLOTTING DEVICES, ARRAYS AND HEADINGS
 7    C
 8    C    TITLE(60) IS THE HEADING FOR PRINTED AND PLOTTED OUTPUT
 9    C    CODE THE CHARACTERS IN BLOCKS OF ONE AND FILL UP
10    C    THE REMAINING BLOCKS WITH BLANKS
11    C
12    C    VNAME(48) IS THE SET OF LABELS FOR OUTPUT VARIABLES
13    C    CODE THE CHARACTERS IN BLOCKS OF ONE
14    C    EACH VARIABLE CAN HAVE UP TO SIX CHARACTERS
15    C    FILL ALL REMAINING BLOCKS WITH BLANKS
16    C    V(8) IS AN ARRAY FOR THE OUTPUT VARIABLES
17    C    INDEX(5) CONTAINS INDICES FOR OUTPUT VARIABLES TO
18    C    BE PLOTTED--USE ZEROS TO FILL UP THE ARRAY
19    C
20    C         COMPUTER SYSTEM CONSTANTS
21    C
22    C    MOUTV IS INTEGER NO. OF OUTPUT DATA SET REF. NO.
23    C    MINV IS INTEGER NO. OF INPUT DATA SET REF. NO.
24    C    LINEPP IS NUMBER OF LINES ON PRINTOUT PAGE
25    C    NCOLS IS NUMBER OF COLUMNS ON OUTPUT DEVICE
26    C    BIGPOS IS LARGEST REAL POSITIVE NUMBER
27    C    BIGNEG IS LARGEST REAL NEGATIVE NUMBER
28    C------------------------------------------------------------
29          COMMON/SYSCON/MOUTV,MINV,LINEPP,NCOLS,BIGPOS,BIGNEG
30          DIMENSION TITLE(60),VNAME(48),V(8),INDEX(5)
31          DATA TITLE/'A','N','I','M','A','L',' ','P',
32         +'O','P','U','L','A','T','I','O','N',' ',
33         +'M','O','D','E','L',37*' '/
34          DATA VNAME/'Y','F',' ',' ',' ',' ',' ','A','F',
35         +' ',' ',' ',' ',' ','T','O','T','S','F',' ',
36         +'Y','M',' ',' ',' ',' ',' ','A','M',' ',' ',' ',
37         +' ','T','O','T','S','M',13*' '/
38          DATA IX/1162261467/
39    C     SET VARIABLES FOR COMPUTER SYSTEM
40          MOUTV=6
41          MINV=5
42          LINEPP=48
43          NCOLS=120
44          BIGPOS=1.0E75
45          BIGNEG=-1.0E75
46          N=0
47    C     INITIALISATION+ AND DATA INPUTS
48          DATA YF,AF,YM,AM,TOTSF,TOTSM/70.,200.,60.,55.,0.,0./
49          DATA PRNF,PRNM,PSLF,PSLM/0.52,0.48,0.29,1.0/
50          REAL TPBR(8),TPDRF(8),TPDRM(8)
51          DATA TPBR/0.,0.45,0.25,0.55,0.75,0.65,1.,0.75/
52          DATA TPDRF/0.,0.,0.25,0.04,0.75,0.06,1.,0.1/
53          DATA TPDRM/0.,0.02,0.25,0.06,0.75,0.08,1.,0.12/
54    C     TIME LOOP SET TO NUMBER OF CALCULATION INTERVALS
55    C     IN TOTAL RUN TIME
56          DO 300 ITIME=1,50
57    C     FLOWS
58    C         BIRTHS
59          RNUM1=RNDX(IX)
60          PBR=TABFN(TPBR,4,RNUM1)
61          OBR=AF*PBR
62          BRF=OBR*PRNF
63          BRM=OBR*PRNM
64    C         DEATHS
65          RNUM2=RNDX(IX)
66          PDRF=TABFN(TPDRF,4,RNUM2)
```

It should be noted that functions for generating pseudo-random numbers are typically machine-dependent. While the function RNDX is included in the listing given above for the sake of completeness, the user should seek expert advice on an appropriate form for use on his particular machine. Fortunately, the use of such generators is so common that they are normally readily available in a suitable form for simple

```
67                    DRF=YF*PDRF
68                    RNUM3=RNDX(IX)
69                    PDRM=TABFN(TPDRM,4,RNUM3)
70                    DRM=YM*PDRM
71        C              SURVIVAL TO MATURITY
72                    SRF=YF*1.0-PDRF
73                    SRM=YM*1.0-PDRM
74        C              SALES
75                    SLRF=AF*PSLF
76                    SLRM=AM*PSLM
77        C              COMPARTMENTS
78                    YF=YF+BRF-DRF-SRF
79                    YM=YM+BRM-DRM-SRM
80                    AF=AF+SRF-SLRF
81                    AM=AM+SRM-SLRM
82                    TOTSF=TOTSF+SLRF
83                    TOTSM=TOTSM+SLRM
84        C    STORE OUTPUT VARIABLES IN V(8)
85                       V(1)=YF
86                       V(2)=AF
87                       V(3)=TOTSF
88                       V(4)=YM
89                       V(5)=AM
90                       V(6)=TOTSM
91        C    PRINT OUTPUT VARIABLES
92                    CALL PRINT(V,ITIME,TITLE,VNAME,6)
93        C    END OF TIME LOOP
94           300 CONTINUE
95                    STOP
96                    END
```

```
C------------------------------------------------------------------
C   THIS FUNCTION RETURNS NUMBER UNIFORMLY DISTRIBUTED BETWEEN 0
C      AND 1
C------------------------------------------------------------------
        FUNCTION RNDX(IX)
        IX=IX*1162261467
        IF(IX.LT.0)IX=IX+2147483647+1
        RNDX=IX*0.4656612E-9
        RETURN
        END
```

applications as in this example. For information on the principles of their construction and special applications see Davies (1971).

Comparing the output in Figs.3.7 (p.65) and A2.5, it will be seen that the numerical values differ, resulting from the use of different pseudo-random number generators in the FORTRAN and CSMP versions.

APPENDIX II

ANIMAL POPULATION MODEL

TIME	YF	AF	TOTSF	YM	AM	TOTSM
1	67.031	211.95	58.000	61.182	59.938	55.000
2	62.463	217.42	119.47	56.687	61.065	114.94
3	48.130	216.77	182.52	41.238	56.573	176.00
4	77.474	202.02	245.38	69.603	41.166	232.58
5	58.942	220.86	303.97	52.698	69.535	273.74
6	77.880	215.75	368.02	68.333	52.621	343.28
7	65.801	231.02	430.58	59.055	68.262	395.90
8	67.697	229.77	497.58	61.277	58.985	464.16
9	59.422	230.77	564.21	54.352	61.201	523.14
10	72.555	223.26	631.14	64.898	54.305	584.34
11	50.195	231.02	695.88	45.430	64.827	638.65
12	55.452	214.20	762.88	50.414	45.396	703.48
13	78.707	207.48	824.99	73.891	50.389	748.87
14	71.237	225.97	885.16	64.589	73.827	799.26
15	69.608	231.64	950.70	65.512	64.565	873.09
16	69.998	234.03	1017.9	63.155	65.448	937.65
17	86.094	236.13	1085.7	77.255	63.096	1003.1
18	70.293	253.73	1154.2	58.845	77.154	1066.2
19	81.091	250.36	1227.8	75.489	58.767	1143.4
20	75.894	258.78	1300.4	69.531	75.421	1202.1
21	96.107	259.58	1375.4	86.924	69.457	1277.5
22	76.680	280.36	1450.7	71.487	86.883	1347.0
23	64.146	275.68	1532.0	58.656	71.425	1433.9
24	86.981	259.86	1612.0	75.546	58.552	1505.3
25	65.752	271.43	1687.3	59.859	75.483	1563.9
26	77.104	258.42	1766.1	70.760	59.798	1639.3
27	70.230	260.49	1841.0	67.529	70.707	1699.1
28	68.233	255.12	1916.5	60.368	67.434	1769.8
29	75.689	249.34	1990.5	66.999	60.293	1837.3
30	91.985	252.66	2062.8	87.476	66.976	1897.6
31	67.243	271.31	2136.1	64.709	87.442	1964.5
32	83.604	259.87	2214.8	72.760	64.638	2052.0
33	86.693	268.10	2290.1	73.197	72.651	2116.6
34	56.769	276.96	2367.9	54.167	73.132	2189.3
35	76.629	253.37	2448.2	69.367	54.096	2262.4
36	69.345	256.50	2521.7	61.011	69.304	2316.5
37	85.677	251.40	2596.1	80.992	60.981	2385.8
38	75.266	264.12	2669.0	68.078	80.923	2446.8
39	88.038	262.78	2745.6	75.880	67.985	2527.7
40	83.052	274.61	2821.8	73.651	75.838	2595.7
41	76.139	277.99	2901.4	70.256	73.612	2671.5
42	101.23	273.45	2982.0	91.876	70.178	2745.1
43	94.398	295.34	3061.3	84.159	91.800	2815.3
44	92.437	304.04	3147.0	84.201	84.087	2907.1
45	99.650	308.26	3235.2	89.802	84.133	2991.2
46	92.342	318.51	3324.5	83.381	89.779	3075.3
47	118.26	318.44	3416.9	107.15	83.311	3165.1
48	85.099	344.26	3509.3	81.458	107.08	3248.4
49	116.37	329.43	3609.1	109.22	81.391	3355.5
50	70.813	350.17	3704.6	66.960	109.14	3436.9

Fig.A2.5. Tabular output from the animal population model.

REFERENCES

CHAPTER 1

Text references

Jeffers, J.N.R. (1974). Future prospects of systems analysis in ecology. *Proceedings of the First International Congress of Ecology*, pp. 255-9. PUDOC, Wageningen, The Netherlands.

Innis, G.S. (1975). The use of a systems approach in biological research. In *Study of agricultural systems* (G.E.Dalton (Ed.)), pp. 369-91. Applied Science Publishers, London.

Radford, P.J. (1968). Systems, models and simulation. In *Annual report of the Grassland Research Institute, 1967*, pp. 77-85. GRI Hurley, Berkshire.

Spedding, C.R.W. (1975). The study of agricultural systems. In *Study of agricultural systems* (G.E. Dalton (Ed.)), pp. 3-19. Applied Science Publishers, London.

────── and Brockington, N.R. (1976). Experimentation in agricultural systems. *Agricultural systems* 1, 47-56.

de Wit, C.T., Brouwer, R., and Penning de Vries, F.W.T. (1970). The simulation of photosynthetic systems. In *Prediction and measurement of photosynthetic activity. Proceedings of International Biological Programme Technical Meeting, Trebon*, PUDOC, Wageningen, The Netherlands.

────── and Goudriaan (1974). *Simulation of ecological processes*. PUDOC, Wageningen, The Netherlands.

Suggestions for further reading

Forrester, J.W. (1961). *Industrial dynamics*. M.I.T. Press, Cambridge, Massachusetts. (Chapters 1-5.)

Novikoff, A.B. (1945). The concept of integrative levels and biology. *Science* 101, 209-15.

Shannon, R.W. (1975). *System simulation: the art and science*. Prentice Hall, New York.

CHAPTER 2

Forrester, J.W. (1961). *Industrial dynamics*. M.I.T. Press, Cambridge, Massachusetts. (Chapter 6 *et seq.*)

Spedding, C.R.W. (1975). *The biology of agricultural systems*. Academic Press, London. (Chapters 2 and 3.)

CHAPTER 3

Text references

Brennan, R.D., de Wit, C.T., Williams, W.A., and Quattrin, E.V. (1970). The utility of a digital simulation language for ecological modelling. *Oecologia (Berlin)* $\underline{4}$, 113-32.

Charlton, P.J. (1971). Computer languages for system simulation. In *Systems analysis in agricultural management*. (J.B. Dent and J.R. Anderson (Eds.)) pp.53-70. J. Wiley, Sydney, Australia.

Davies, R.G. (1971). *Computer programming in quantitative biology*. Academic Press, London.

Frissel, M.J. and Reiniger, P. (1974). *Simulation of accumulation and leaching in soils*. PUDOC, Wageningen, The Netherlands.

Naylor, T.H., Balintfy, J.L., Burdick, D.S., and Chu, K. (1966). *Computer simulation techniques*. J.Wiley, New York.

Radford, P.J. (1972). The simulation language as an aid to ecological modelling. In *Mathematical models in ecology: Twelth Symposium of the British Ecological Society*. (J.N.R. Jeffers (Ed.)), pp.277-96. Blackwell, Oxford.

de Wit, C.T. and van Keulen, H. (1972). *Simulation of transport processes in soils*. PUDOC, Wageningen, The Netherlands.

——— and Goudriaan, J. (1974). *Simulation of ecological processes*. PUDOC, Wageningen, The Netherlands.

Texts on computer programming and reference manuals

IBM Program Number 5734-XS9 (1975). *Continuous System Modelling Program (CSMP III), Program reference manual.*

McCracken, D.D. (1970). *A guide to FORTRAN programming* (2nd ed.). J. Wiley, New York.

Organick, E.I. (1966). *A FORTRAN IV primer*. Addison-Wesley, Reading, Massachusetts.

Pugh, A.L. (1970). *DYNAMO II users' manual*. M.I.T. Press, Cambridge, Massachusetts.

Further reading

Forrester, J.W. (1961). Industrial dynamics. M.I.T. Press, Cambridge, Massachusetts. (Chapters 6-8.)

Milne, E.W. (1960). *Numerical solution of differential equations*. McGraw-Hill, New York.

Patten, B.C. (Ed.) (1971). *Systems analysis and simulation in ecology*. Vol.1. Academic Press, New York. (Chapters 1 and 2.)

See also Davies (1971), Chapters 1-3; de Wit and Goudriaan (1974), Chapters 2 and 3.

CHAPTER 4

Text references

Blackman, F.F. (1905). Optima and limiting factors. *Annals of Botany* 19, 281.

Brockington, N.R. (1971). Using models in agricultural research. *Span*, 14, 26-9.

——————————— (1972). A mathematical model of pasture contamination by grazing cattle and the effects on herbage intake. *J.Agricultural Science (Camb.)*. 79, 249-57.

Frissel, M.J. and Reiniger, P. (1974). *Simulation of accumulation and leaching in soils*. PUDOC, Wageningen, The Netherlands.

Forrester, J.W. (1961). *Industrial dynamics*. M.I.T. Press, Cambridge, Massachusetts.

——————————— (1968). *System dynamics*. M.I.T. Press, Cambridge, Massachusetts.

Hillel, D. (1977). *Computer simulation of soil-water dynamics*. Monograph of International Development Research Centre, Ottawa, Canada.

van Keulen, H. (1975). *Simulation of water use and herbage growth in arid regions*. PUDOC, Wageningen, The Netherlands.

Liebig, J. (1840). *Die organische Chemie in ihrer Anwendung auf Agricultur und Physiologie*. Viewig, Brausnchweig.

Rabinowitch, E.I. (1951). *Photosynthesis and related processes*. Vol.II, Pt.1. Interscience Publishers, New York.

Radford, P.J. and Greenwood, D.J. (1970). The simulation of gaseous diffusion in soils. *J.Soil Science* 21, 304-13.

Ryle, G.J.A., Brockington, N.R., Powell, C.E., and Cross, B. (1973). The measurement and prediction of organ growth in barley. *Annals of Botany* 37, 233-46.

de Wit, C.T. and van Keulen, H. (1972). *Simulation of transport processes in soils*. PUDOC, Wageningen, The Netherlands.

de Wit, C.T. and Goudriaan, J. (1974). *Simulation of ecological processes*. PUDOC, Wageningen, The Netherlands.

Suggestions for further reading

(A) The following general references are an extension of the lists given in previous chapters.

Arnold, G.W. and de Wit, C.T. (Eds.) (1976). *Critical evaluation of systems analysis in ecosystems research and management*. PUDOC, Wageningen, The Netherlands.

Dalton, G.E. (Ed.) (1975). *Study of agricultural systems*. Applied Science Publishers, London.

Dent, J.B. and Anderson, J.R. (1971). Systems analysis in agricultural management. J. Wiley, Sydney, Australia.

Jeffers, J.N.R. (Ed.) (1972). *Mathematical models in ecology. Twelth Symposium of the British Ecological Society*. Blackwell, Oxford.

Patten, B.C. (Ed.) (1972). *Systems analysis and simulation in ecology*. Vol.II. Academic Press, New York.

————   (1975). *Systems analysis and simulation in ecology*. Vol.III. Academic Press, New York.

Prediction and measurement of photosynthetic activity. *Proceedings of the International Biological Programme Plant Physiology Technical Meeting, Trebon (1970)*. PUDOC, Wageningen, The Netherlands.

Thornley, J.H.M. (1976). *Mathematical models in plant physiology*. Academic Press, London.

Watt, K.E.F. (Ed.) (1966). *Systems analysis in ecology*. Academic Press, New York.

(B) Journals of particular relevance to simulation modelling in agriculture include:

Agricultural Systems. Applied Science Publishers Ltd., London.

Ecological Modelling. Elsevier Scientific Publishing Co., Amsterdam, The Netherlands.

Simulation. Technical Journal of the Society for Computer Simulation, La Jolla, California, USA.

It should be noted, also, that relevant papers frequently appear in journals devoted to the biological/agricultural *subjects* of particular modelling exercises.

CHAPTER 5

Text references

Baker, C.T. and Dzielinski, B.P. (1960). Simulation of a sim-
    plified job shop. *Management Science*, 6, 311-23.

Blake, J. and Gordon, G. (1964). Systems simulation with a
    digital computer. *IBM Systems Journal* 3, 14-20.

Conway, R.W., Johnson, B.M., and Maxwell, W.L. (1959). Some
    problems of digital systems simulation. *Management
    Science* 5, 92-110.

Edelsten P.R. (1976). A stochastic model of the weather at
    Hurley in S.E. England. *Meteorological Magazine* 105,
    206-14.

Efron, R. and Gordon, G. (1964). A general purpose systems
    simulator and examples of its application. *IBM Systems
    Journal* 3, 22-34.

Feyerham, R. and Bark, L.D. (1967). Goodness of fit of a
    Markov chain model for sequences of wet and dry days.
    J. of Applied Meteorology 6, 770-3.

Geisler, P.A., Paine, A.C., and Geytenbeek, P.E. (1977).
    Simulation of an intensified lambing system incorpora-
    ting two flocks and the rapid re-mating of ewes.
    Agricultural Systems 2, 109-20.

Holling, C.S. (1966). The functional response of inverte-
    brate predators to prey density. *Memoirs of Entomological
    Society of Canada* 48, 1-86.

James, A.D. (1977). Models of animal health problems. *Agri-
    cultural Systems* 2, 193-7.

Krasnow, H.S. and Merikallio, R.A. (1964). The past, present
    and future of general simulation languages. *Management
    Science* 11, 236-67.

Leslie, P.H. (1945). On the use of matrices in certain popu-
    lation mathematics. *Biometrika* 35, 183-212.

Tocher, K.D. (1965). Review of simulation languages. *Opera-
    tional Research Quarterly* 16, 105-34.

Usher, M.B. (1972). Developments in the Leslie Matrix Model.
    In *Mathematical models in ecology* (Jeffers, J.N.R. (Ed.)
    Twelfth Symposium of the British Ecological Society.
    Blackwell, Oxford.

Event-oriented simulation languages

I.B.M. (1971). General Purpose Simulation System V Users
    Manual (SH20-0851-1). *Introductory users' manual*
    (SH20-0866).

Hanser, R.B., Markowitz, H.M., and Karr, H.W. (1962). *SIMSCRIPT - a simulation programming language*. Rand Corporation, RM 3310.

Xerox Corporation (1972). Xerox General Purpose Discrete Simulator. *Reference manual* (90-17-583).

CHAPTER 6

Text references

Amidon, E.L. and Akin, G.S. (1968). Dynamic programming to determine optimum levels of growing stock. *Forest Science* 14, 287-9.

Arnold, G.W. and de Wit, C.T. (Eds.) (1976). *Critical evaluation of systems analysis in ecological research*. PUDOC, Wageningen, The Netherlands.

Baldwin, R.L., Koong, L.J., and Ulyatt, M.J. (1977). A dynamic model of ruminant digestion for evaluation of factors affecting nutritive value. *Agricultural Systems* 2, 255-88.

Bellman, R.E. (1957). *Dynamic Programming*. Princeton University Press, New Jersey.

Crabtree, J.R. (1972). The development of a milk production model. *Oxford Agrarian Studies* 1, 83-96.

Dale, M.B. (1970). Systems analysis and ecology. *Ecology* 51, 1-16.

Dent, J.B. and Casey, H. (1967). *Linear programming and animal nutrition*. Crosby Lockwood, London.

Edelsten, P.R. and Newton, J.E. (1977). A simulation model of a lowland sheep system. *Agricultural Systems* 2, 17-32.

Flinn, J.C. and Musgrave, W.F. (1967). Development and analysis of input/output relations for irrigation water. *Australian J. of Agricultural Economics* 11, 1-19.

Heady, E.O. and Candler, W. (1968). *Linear programming methods*. Iowa State University Press, Ames, Iowa.

Innis, G.S. (1973). The use of a systems approach in biological research. In *Study of agricultural systems*, (Dalton, G.E. (Ed.)), pp. 389-91. Applied Science Publishers, London.

Jacobs, O.L.R. (1967). *An introduction to dynamic programming - the theory of multi-stage processes*. Chapman Hall, London.

Jeffers, J.N.R. (1976). Future prospects of systems analysis in ecology. In *Critical evaluation of systems analysis in ecosystems research and management.* (Arnold, G.W. and de Wit, C.T. (Eds.)), pp. 98-108, PUDOC, Wageningen, The Netherlands.

Kaufman, A. and Cruon, R. (1967). *Dynamic programming - sequential scientific management.* Academic Press, New York.

van Keulen, H. (1976). Evaluation of models. In *Critical evaluation of systems analysis in ecosystems research and management* (G.W. Arnold and C.T. de Wit (Eds.)), pp.22-9. PUDOC, Wageningen, The Netherlands.

Maxwell, T.J., Eadie, J., and Sibbald, A.R. (1973). Methods of economic appraisal of hill sheep production systems. *Potassium Institute Colloquium Proceedings* No. 3, pp. 103-13.

Miller, D.R., Weidhas, D.W., and Hall, R.E. (1973). Parameter sensitivity in insect population modelling. *J.Theoretical Biology* $\underline{42}$, 263-74.

Naylor, T.H., Balintfy, J.L., Burdick, D.S., and Chu, K. (1966). *Computer simulation techniques.* J. Wiley, New York.

Penning de Vries, F.W.T. (1977). Evaluation of simulation models in agriculture and biology: conclusions of a workshop. *Agricultural Systems* $\underline{2}$, 99-108.

Powell, M.J.D. (1964). An efficient method of finding the minimum of a function of several variables. *Computer J.* $\underline{7}$, 155-61.

Quenouille, M.H. (1957). *Analysis of multiple time series.* Griffin, London.

Radford, P.J. (1972). The simulation language as an aid to ecological modelling. In *Mathematical models in ecology* (Jeffers, J.N.R. (Ed.)), pp. 277-96. Blackwell, Oxford.

Shannon, R.E. (1975). *Systems simulation: the art and science.* Prentice-Hall, New York.

Smith, F.E. (1970). Analysis of ecosystems. In *Analysis of temperate forest ecosystems.* (Reichle, D.E. (Ed.)), pp. 7-18. Springer-Verlag, New York.

Steinhorst, R.K., Hunt, H.W., Innis, G.S., and Haydock, K.P. (1978). Sensitivity analyses of the ELM model. In *Grassland simulation model* (Innis, G.S., (Ed.)), pp. 231-56. Springer-Verlag, New York.

Swartzman, G.L. and Van Dyne, D.M. (1972). An ecologically based simulation-optimization approach to natural resource planning. *Annual Review of Ecology and Systematics* <u>3</u>, 347-98.

Teichroew, D. (1965). A history of distribution sampling prior to the era of the computer and its relevance to simulation. *J. of American Statistical Association* <u>60</u>, 27-49.

Thornley, J.H.M. (1976). *Mathematical models in plant physiology*. Academic Press, London.

Tomovic, R. (1963). *Sensitivity analysis of dynamic systems*. McGraw-Hill, New York.

Tomovic, R. and Vukobratovic, M. (1970). *General sensitivity theory*. Elsevier, New York.

de Wit, C.T., Brouwer, R., and Penning de Vries, F.W.T. (1970). The simulation of photosynthetic systems. In *Prediction and measurement of photosynthetic activity. Proceedings of the International Biological Programme Plant Physiology Technical Meeting, Trebon*, pp. 47-70. PUDOC, Wageningen, The Netherlands.

————   and van Keulen, H. (1972). *Simulation of transport processes in soils*. PUDOC, Wageningen, The Netherlands.

Wright, A. (1971). Farming systems, models and simulation. In *Systems analysis in agricultural management* (Dent, J.B. and Anderson, J.R. (Eds.)), pp. 17-34. J. Wiley, Sydney, Australia.

Linear programming texts

Cooper, W.C. and Charnes, A. (1968). Linear programming. In *Mathematics in the modern world; readings from the Scientific American*. Freeman, San Francisco.

Dantzig, G.B. (1963). *Linear programming and extensions*. Princeton University Press, New Jersey.

Hadley, G. (1963). *Linear programming*. Addison-Wesley, Reading, Massachusetts.

Orchard-Hays, W. (1968). *Advanced linear programming computing techniques*. McGraw-Hill, New York.

AUTHOR INDEX

Page numbers underlined indicate reference citations in full.

SUBJECT INDEX